To aid you in writing successful grant proposals,
samples of successful proposals and a directory of useful links
are available FREE on-line.

To view these materials,
please visit

www.josseybass.com/go/sciencegrants

Thank you,
Thomas R. Blackburn

Getting Science Grants

Getting Science Grants

Effective Strategies for Funding Success

Thomas R. Blackburn

JOSSEY-BASS
A Wiley Imprint
www.josseybass.com

Published by Jossey-Bass
A Wiley Imprint
989 Market Street, San Francisco, CA 94103-1741   www.josseybass.com

Jossey-Bass books and products are available through most bookstores. To contact Jossey-Bass directly, call our Customer Care Department within the U.S. at (800) 956-7739, outside the U.S. at (317) 572-3993 or fax (317) 572-4002.

Jossey-Bass also publishes its books in a variety of electronic formats. Some content that appears in print may not be available in electronic books.

Library of Congress Cataloging-in-Publication Data
Blackburn, Thomas R., 1936–
    Getting science grants : effective strategies for funding success /
Thomas R. Blackburn—1st ed.
        p. cm.
Includes bibliographical references and index.
    ISBN 0-7879-6746-7 (pbk.)
    1. Research grants. I. Title.
    Q180.55.G7B53 2003
    507'.9—dc21   2003013893

Printed in the United States of America
FIRST EDITION
PB Printing   10 9 8 7 6 5 4 3 2 1

Contents

Preface

Every year, tens of thousands of scientists worldwide receive letters that begin with the words, "We regret to inform you that your proposal was not recommended for funding." The letter then goes on to say that the number of very good proposals submitted was more than the available funds could support. Kind thoughts with regard to the scientist's interest in the fund and wishes for success in obtaining funding for the spurned research are often expressed.

If that scientist is you, you might sigh, or swear, file the letter in the trash, and then pass on the bad news to the students and postdoctoral fellows who will not be supported by that grant. Then you get to work on the next proposal. This book will take you beyond kind thoughts and wishes and will help you succeed in funding your research and teaching.

Scientists share a need for research sponsorship, whether they are at traditional research universities, branch campuses, state colleges, or private liberal arts colleges. It has become widely recognized that really learning science is almost indistinguishable from doing it. Thus, scientific research has a central role not only in the training of future university faculty and research professionals, but also in the education of future teachers, businesspeople, entrepreneurs, and citizens.

But doing science is expensive, and it is the responsibility of faculty to bring in most of the funding needed, generally from sources outside their institutions. In the United States, most funding is awarded competitively, based on some form of expert, usually anonymous, peer review of proposals. The ability to write convincing, competitive proposals is therefore an indispensable tool in the kit of every scientist. But as important as this is, few colleges and universities teach this skill.

Rather, it is learned, if at all, through trial and error and frustration. This book was written to help you acquire proposal writing skills and use them effectively to get research and curricular development money from agencies and foundations in the United States.

Consistent success in funding requires good ideas, presented in excellent proposals. Anything less will result in lost time and effort. In this book, I assume that the reader will supply the good idea. It is my goal to make sure it is presented so that it has its best chance of obtaining funding.

I and the many people I have talked to about grants have, in aggregate, many decades of experience in college and university teaching and research and as program officers at major scientific funding agencies. I have written my share of successful research and curricular proposals and have read or reviewed thousands more written by scientists from around the world. I have solicited and read more than ten thousand anonymous peer reviews of those proposals, sitting in on innumerable panel meetings where the proposals are discussed, critiqued, and ranked for funding by leading scientists.

What This Book Offers

This book is written to be useful to scientists across many fields of natural science, including physics, chemistry, geology, psychology, astronomy, biology, materials science, and engineering, as well as hybrid fields like biomedicine and planetary science. It will be helpful in writing grant proposals to the many agencies, both public and private, that support scientific research and curricular development in all fields. In particular, this book targets college and university faculty, postdoctoral fellows, and graduate students who are now, or soon will be, responsible for funding their own research and teaching.

On the companion Web site for this book, www.josseybass.com/go/science-grants, you will find links to a wide variety of funding sources for scientists and the actual text of winning proposals in biology, chemistry, psychology, geology, planetary science, physics, and other fields that resulted in grants from a variety of agencies and grant programs, both public and private (see Resource B for a complete listing of the contents of the Web site).

This book offers tools that will allow you to

- Get a realistic understanding of the grant process
- Give your research ideas clear and compelling definition
- Choose the funding agencies where you have the best chance of winning support
- Write each section of a proposal clearly and persuasively
- Recognize and avoid common mistakes in proposal writing and relations with funders
- Recognize the characteristics of excellent proposals and incorporate them into yours
- Profit from the experience of being turned down and reduce the chance that it will happen again
- Manage success so as to continue it

Some of the advice you will find here may strike you as bare common sense parading in the emperor's clothing of expertise. You will be right: quite a lot of the advice in this book is the result of seeing what happened when well-trained scientists, including me, forgot to use common sense. The rest consists of lore that you will be better off learning here than by trial and error.

There is a body of practices, ethics, and folkways that governs all the major granting agencies. I have tried to lead readers to a good understanding of these common attitudes and practices. However, details of agency practices and funding criteria vary substantially from one agency to another. Because I want this book to be useful for writers of applications to all of those agencies, I have had to use hedge wording such as *often, usually, virtually always,* and so forth when describing agency practices. Absolutist language would be possible only if I were to confine myself to describing the practices of a single grant program of a single agency. I have tried to be judicious in using these terms and not to say *usually* when I mean *often* and (almost) never saying *never*.

At several points, I refer to a particular commercial service such as the Google search engine, the U.S. Postal Service, or the IRIS grants database. These (and many other such references) are intended as examples only; in no case does the mention of a commercial service constitute an endorsement of its products, price structure, or results.

Acknowledgments

I would like to acknowledge colleagues and fellow program officers across the science funding arena for invaluable discussions and for their collective advice and tips for up-and-coming as well as experienced proposal writers, extending over five decades of involvement in grants. In particular, I thank Andy Colb, Justin Collat, Lawrence Funke, John Malin, Barbara Ransom, Robert Rich, Joseph Rogers, and Ronald Siatkowski of the American Chemical Society Petroleum Research Fund; Robert Lichter of the Camille and Henry Dreyfus Foundation; Catherine T. Sigal of the Juvenile Diabetes Research Foundation; Janet E. Nelson and Alan Willard of the National Institutes of Health; Henry Blount, Jody Bourgeois, Chris Maples, and Jill Singer of the National Science Foundation; Brian Andreen, Ray Kellman, and Scott Pyron of the Research Corporation; and Elaine Hoagland of the Council on Undergraduate Research. I particularly thank Richard Ramette of Carleton College, who was the principal investigator on the grant that supported my first toddling steps in scientific research.

Chris Arumainayagam of Wellesley College, Erik Jorgensen of the University of Utah, Catherine Jahncke of St. Lawrence University, Miriam Kastner of Scripps Oceanographic Institution, Ian Macdonald of Texas A&M University, Scott McLennan of the State University of New York at Stony Brook, Maria Ngu-Schwemlein of the University of South Alabama, Gary Radice of the University of Richmond, Janet Stiles of the University of California at San Diego, Martha Stipanuk and Patrick J. Stover of Cornell University, and MingMing Wu of Occidental College provided invaluable consultation and proposal excerpts for the companion Web site. Members of the Vicksbury Blackfryars read several chapters and made numerous valuable suggestions. I thank Skip Hadley for help with PDF files. I particularly thank Barbara Ransom, without whose encouragement this book would not have been written.

Any errors of fact and all wrong-headed opinions are entirely my responsibility.

Washington, D.C.   Thomas R. Blackburn
May 2003

The Author

Thomas R. Blackburn holds undergraduate and doctoral degrees from, respectively, Carleton College and Harvard University. After thirty years as a faculty member at several liberal arts colleges, he joined the American Chemical Society Petroleum Research Fund (ACS PRF), serving for ten years as assistant program administrator and senior program officer. He is the author of *Equilibrium: A Chemistry of Solutions* and coauthor of *Chemistry: Molecules That Matter*.

His teaching and research in analytical chemistry and geochemistry have been supported by grants from the National Science Foundation, the National Aeronautics and Space Administration, the ACS PRF, Harvard University, and the governments of Great Britain and Switzerland. He lives in Washington, D.C., where he is a member of the Geochemical Society, the American Association for the Advancement of Science, the Geological Society of Washington, and the Hunt String Quartet.

Getting Science Grants

Planning for Funding Success

This chapter lays the groundwork for you to write excellent, competitive proposals for grants to support your research. It surveys the work before you and the journey your proposal will travel after you finish it. The chapter also gives you a first look at many ideas that are more fully developed later in the book.

This chapter introduces the three key supports of a winning proposal: (1) a thorough survey of the state of affairs in your research area, (2) a clear description of how you plan to contribute to it, and (3) a convincing discussion of why your contribution is important and should be funded. In the process, I will offer some observations about not promising too much or too little, about working with a co-investigator, and other issues that bear on the central task of shaping successful research proposals.

Very few people—they say Mozart was one—can create a completed masterpiece in their heads. Proposal writing, like all creative work, is an iterative process. You simply cannot accomplish a finished proposal in a single, or even third, draft. Rather, writing and thinking are intertwined. The act of writing jogs your mind toward new ideas and approaches, and as you work out these ideas, you change what you have already written.

It may appear that you need to complete all kinds of preliminary steps before putting the first word of your proposal on paper, but that is not so! The preliminaries are shaped by the content of the proposal, and that changes as you write. For the sake of an orderly book, I begin with the preliminaries and then move on to the proposal itself, but in actuality, your writing should begin as soon as you have the first inkling of an idea. As you write, you will develop your ideas, and your finished proposal will be vastly more focused, sound, interesting, and competitive than the one you started with.

A Pep Talk

You might be thinking at this point, *How level is this playing field, anyhow?* Some of what I say in this book may strike you as idealistic, ignoring the politics of big-time, big-dollar science. We've all heard the grumblings: "They only fund their buddies" or, "No one at this institution ever has gotten, or will get, a grant from the Gigabux Foundation" or the corollary, "When you're famous enough, even your bad proposals get funded."

My experience, and that of other program officers whose opinion I trust, does not bear out such self-fulfilling pessimism. It is true that proposal writers at institutions that have not yet established a reputation for scientific research have an extra burden of proof. I will address that situation in Chapter Five. Otherwise, what I have found is that those fallible and political human beings who read, review, and recommend research proposals for funding nearly always try to support the most promising and exciting science they see, regardless of who proposes it or where it comes from. And they do so on the basis of straightforward and transparent principles that anyone who has studied science can readily understand and apply. This book exists to help you do just that. Let's start with a road map.

The Four Stages of Proposal Review: A Preliminary Map

Every scientific research proposal encounters four stages of evaluation before its fate—funding or denial—is final: (1) initial review by the managers (usually called program officers) who staff the funding agency; (2) technical peer review; (3) ranking by an agency panel of advisory experts or by the program officer, combined with recommendation for funding or denial; and (4) final approval of the recommended list of grants.

This book treats each of these four functions as a separate step, done by different people or by the same people wearing different hats. And for many funders, that is exactly the case. For other agencies, these four functions may overlap and combine in various ways. For example, at some agencies, program officers have authority to rank and recommend funding or denial; at others, that power is reserved to the panel. Some agencies, or some programs within agencies, may do without the panel, giving all but the final approval authority to the program officer. But all four functions—program officer review, technical peer review, ranking (with recommendation for funding or denial), and final approval—are there every time in some form. Let's survey each of them in turn.

Program Officer Review

Part of the job of the program officer is to make sure that the valuable time of the technical experts who will evaluate a proposal is not wasted in reading those that have no chance of being funded because they violate some agency guideline (and, equally, to prevent your wasting your own time in writing a doomed proposal). I will talk more about these guidelines later, but a single example will show you what I mean.

Some grants programs target research in undergraduate colleges. A proposal to such a program from the physics department of a major graduate university would not move to the next level of review, no matter how good the science it proposed. And it is the program officer's job, in general, to make sure that all of the proposals that go on to the next level of review meet all of the eligibility requirements and other guidelines. Because there are a lot more proposals than there are program officers, this step can take from a few weeks to a month or more.

You can be sure of passing this first level of review only by contacting the agency before you submit and making sure that your project, your institution, and you yourself are eligible for a grant under the guidelines of the program you are applying to. A simple telephone call should take care of this, but many, many applicants neglect to make that call. I will have a lot more to say about this in Chapter Two.

Technical Peer Review

This is the step most people have in mind when they talk about having their proposal "reviewed." After the agency receives your proposal, logs it in, assigns it a serial number, and reviews it for adherence to guidelines, the program officer sends copies for evaluation to trusted experts in the field of the research. These may be scientists scattered across the country and abroad who read the proposal and return a review. In some agencies, they will be scientists who gather in one place (often, in the environs of Washington, D.C.) to constitute an ad hoc committee to read and discuss your proposal (and all the others competing with it for funding) before producing a review. In either case, these are people in a good position to make helpful comments on the science and technical issues in the proposal, and their comments are almost always very valuable to you, whether your proposal is ultimately funded or denied funding.

Usually, though not always, you will not know the names or locations of these peer reviewers. Whether expert peer review is more reliable when it is anonymous is not a settled question. My own opinion is that it is better if you do not know the identity of those who will examine your science. It forces you to concentrate on the science itself as you write (rather than a particular person's interests and opinions), and it avoids awkward relations with the reviewer if your proposal is denied funding. Virtually all agencies will show you the reviews that your proposal receives; it is easier to read them objectively if you don't know who wrote them. I have more to say about writing effectively for these nameless readers in Chapters Three through Five.

It almost never takes less than four months, and it often takes more, to get adequate technical peer review of all the proposals that will compete in a group. (I will discuss proposal groups later in this chapter.) These reviews (rarely fewer than three or more than six) are collected by the program officer in preparation for the next step.

Ranking and Recommendation

The defining moment in the life of a proposal comes when it is ranked against all other competing proposals by the panel of scientists or the program officer responsible for recommending funding. In the case of agencies with a disease-related mission, a lay review board may be involved at this stage, to ensure that the grants given meet an interested public's goals for the agency.

In this step, the results of the first two evaluations are combined: every agency wants to fund only the very best science consistent with the aims of the particular grant program. These decisions are usually made during the course of a one- or two-day panel meeting. Thus, both the results of the technical peer review (step 2) and a separate judgment as to how well the proposed work advances the goals of the agency in creating the grant program (step 1) have to be weighed and combined. I will discuss how this feat is accomplished later in this chapter.

Unlike external peer review discussed above, review panels, or "study sections" in National Institutes of Health (NIH) language, are not necessarily anonymous. Agencies often publish the names and institutional affiliations of their panel members in their literature or on their Web site. It is always a good idea, once you have decided on an agency to send a proposal to, to look them up. Get an idea of the kind of work they do themselves and anything else you can learn about them. Because it is a little hard to find on the sprawling NIH site, a link to NIH Study Section rosters is given on the companion Web site for this book (see the document called "A Directory of Useful Information on Grants and Granting Agencies in Science" at www.josseybass.com/go/sciencegrants).

Final Approval of Funding

Grants are almost never made to individual scientists; they are agreements between the funding agency and your institution, providing for certain research to be carried out by you and your students, funded by the institution using money from the agency. (This distinction almost never makes a big difference, but some cases in which it does are discussed in Chapter Seven.)

For most agencies, the legal authority to establish that funding agreement with your college or university resides outside the office, and its advisory panel, that administers the grant program. Thus, the rankings that the panel makes in step 3 do not produce funding *decisions*. They produce *recommendations* that a higher authority (a responsible administrator, a grants committee, a board of directors) has the theoretical power to alter. For very good reasons, having to do with the fact that advisory panels are made up of people who don't want to find that they have wasted their time, it is rare indeed that funding recommendations are reversed at this formal, grantmaking level. For most nongovernmental grant agencies, it has rarely, if ever, happened. When you can get busy scientists to rank and recommend grant proposals, the last thing you want to do is to ignore their advice. However, particularly for federal agencies, either budget limitations or—very rarely—politics may decrease or reverse a funding recommendation.

Thus, no grant is really final until the grantmaking authority signs off on it. This step can add several weeks or months to the process, giving a grand total of from four months to as much as a year between the time you submit the proposal and the time its funding is official. Chapter Six discusses how to spend this time productively.

Scoring, Proposal Quality, and Funding Rates

In the next four chapters, I'll explore how to create a proposal that succeeds in steps 1 and 2, program officer review and technical peer review. Step 4 (final approval) is out of your hands. The heart of the funding process is step 3, ranking and recommendation. The panel review is when proposals are ranked in priority order for funding and grant (or denial) recommendations are made.

All proposals submitted to a given agency or foundation generally do not compete for the same pot of money. Almost always, the competition is limited to one of several subgroups of proposals, defined by the agency. A proposal subgroup might be as specific as "research in synthetic organic chemistry by faculty at undergraduate colleges within three years of their initial appointment" or as broad as "research in high-energy physics."

Those who score and rank proposals commonly assign numerical scores to certain criteria expected to appear in the proposal. For example, out of a possible total of 100 points, a particular proposal might be awarded 15 points out of a maximum of 20 for "Educational Impact." The relative importance of each scored criterion is reflected in the number of points assigned to it, and this depends on agency priorities. Thus, a perfect score of 5 for "Format Conforming to Requirements" can't make up for a poor score for "Educational Impact," at 20 points.

In this, as in many cases to follow, specific weightings and other ranking policies, while typical of agency practices, are only examples. Each agency or competition has its own criteria and point allocation scheme. The NIH uses a priority scoring system, in which the lowest scores, rather than the highest, are recommended for funding. For example, "Priority 1" at NIH is a better ranking than "Priority 2." The scoring principles are analogous, however. (The NIH scoring criteria are set out in more detail in Chapter Three.)

Although evaluators assign these scores as thoughtfully and conscientiously as possible, of course there is no rigorous mathematical connection between the prose of a scientific proposal and the number that is assigned. Also, the members of the agency panel, unlike the technical peer reviewers, may not be experts in exactly your field of science; this could hardly be so, given that the ratio of submitted proposals to panel members is often ten to one or greater. Thus, you must write your proposal so it can be understood not only by experts in your research area, but also by knowledgeable generalists who are ranking and recommending proposals that are outside their immediate expertise. You have to be at the top of the technical game *and* be able to communicate the importance and impact of the proposed work to these generalists.

By weighting and summing the subsection scores, the panel gives your proposal an overall numerical score and then ranks proposals in numerical order of scores. It is from this ranking that recommendations for funding are made.

Technical Reviews and Panel Rankings

The result of technical peer review is virtually always the most important scoring component. Proposals are sent out for peer review weeks to months before they are considered for funding. Reviewers are requested to provide technical comments and a summary score on a scale from "Excellent" through "Very Good," "Good," and "Fair," to "Poor." (Because so many good scientists proposing good science are applying for funding these days—and because of severe inflation in scoring—some agencies have added a superscore such as "Outstanding" or "Truly Exceptional." Reviewers are asked to award these very sparingly.) At some agencies, the average of these external scores weighs very heavily in the final ranking by the panel. At others, the scores are taken as strongly advisory but not controlling. In this book, particularly in Chapter Five, I sometimes use these summary scores as a shorthand for broad classes of proposal characteristics in order to explain how to create excellent ones.

Always bear in mind the difference between your proposal's summary evaluation (on the "Poor" to "Excellent" scale) and its ranking relative to other proposals in its group. Getting a host of "Excellent" scores does not necessarily mean that your proposal will be funded. One of the worst duties a grants officer has is telling the author of a proposal with three "Excellent" reviews out of four that the proposal was not recommended for funding. When this happens, it is generally because that proposal ranked in the second quartile of the group and there was only enough money to fund proposals in the first quartile. I bring this to your attention not to discourage you or demonstrate how hard it is to write a winning proposal but rather to remind you that summary scores are not the whole picture. What summary scores can tell you unequivocally is whether your science is up to snuff or whether it needs additional work.

Tiebreakers

At many agencies, not all proposal quality criteria carry the same weight in making funding recommendations. Among proposals that otherwise meet agency guidelines, perceived scientific quality and promise of impact on future research (discussed in Chapters Three through Five) are the most important. When two or more proposals are ranked very closely in scientific merit and close to the maximum number of proposals that the agency budget will fund, secondary criteria, sometimes called tiebreakers, may become very important. These might include judgments by the panel or the program officer as to the impact that funding, or denial of it, might have on the applicant's laboratory, institution, or students; whether the research represents a new initiative or continuation of an existing line of research for the investigator; or the apparent extent of institutional support for the research (through matching funds or support of students, for example). Agencies generally list their funding priorities in their literature or Web sites. (Those of the National Science Foundation and NIH are provided later in this book.)

How Good Is "Very Good"?

Most program officers agree that the majority of research proposals submitted by college and university scientists describe at least "very good" work that ought to be done and that would advance science if it were done. But there isn't enough money to support all good research. Exhibit 1.1 presents this situation graphically. Imagine all research proposals written in a given year for a given grant program and laid out in a distribution of quality groups, based on peer reviewer scores. This exhibit includes a "Super" category above "Excellent," which is used by some agencies to designate these exceptional proposals.

You can easily appreciate what the problem is when you see that proposals ranked "Poor" are even rarer than the "Super" ones and that the great majority of the proposals are rated "Very Good" or better. Most people would agree that a "Very Good" thing is something that ought to be done. But most grant programs have

Exhibit 1.1. Proposal Quality Distribution

only enough money to fund about 15 to 35 percent of the proposals submitted at any given time (again, a rough average from many agencies with which I am familiar). That fraction of successful proposals, also known as the funding rate, varies from program to program and agency to agency, and from time to time as endowments and funding levels fluctuate.

Your Job: Writing an Excellent Proposal

Your job is to write proposals that will consistently score "Excellent," because that is what it takes to be competitive for funding. Here is one way to break down that task:

- Do not commit a small number of easily avoided mistakes that will put your proposal in the lower third of its group, with scores of "Poor" to "Good." Both new and experienced applicants regularly commit these mistakes.

- Understand and then forever shun a longer list of subtler flaws that can subvert an excellent research idea into a "Very Good" proposal.

- Make sure that your research idea is presented in a convincing manner so that your proposal will earn scores of "Excellent" or better from both the reviewers and the panelists or program officers who read it.

The first two are relatively easy once you recognize the mistakes and writing flaws that lie in wait for the unwary. Chapter Five of this book takes you on an extended tour through that jungle. It is the third task that is the heart of any excellent proposal. Excellence depends on both the intrinsic scientific merit of the work being proposed and the clarity and power with which it is presented. Intrinsic scientific merit, of course, is largely your property, and this book can't supply it, although as you read this book and the funded proposals on the companion Web site, you will see many of its hallmarks. But this book can help you with clarity and power.

A powerfully written proposal stands on three main supports:

- A complete and up-to-date survey of the state of research in the field of your proposal

- A realistic and well-delimited presentation of a well-chosen research idea and of yourself and your institution as the proper place for it to be carried out

- A clear parallel between the impact of the work you propose and the aims of the grants program to which you are applying

Fitting Your Research into the Big Picture

The first question any reviewer will ask about your proposal is, "What's this all about? How does it fit into and advance what I understand to be the state of research in this field?" If you have a newly minted Ph.D. or are a beginning researcher, the chance is good that your picture of the state of your field has been shaped strongly by your previous research mentors, research groups, and institutional seminars. You need to recognize that not everyone shares just this background. There is nothing more embarrassing than submitting a proposal, only to learn from a review that the work you propose has already been done or is nearing completion somewhere else.

Therefore, your first job is to make a thorough, conscientious, and independent search of the published literature and the Internet to be sure that you have a comprehensive grasp of what has been done in the field and what is likely to be done in the next year. Look at the personal and research group Web pages of prominent scientists in your field. Wherever possible, take time to attend seminars and meetings in your area of interest—for example, Gordon Research Conferences (see www.grc.uri.edu), small society meetings, and international symposia. Talking to people at these meetings is one of the best ways to find out quickly what the state of research is, who the main players are, and what they are doing. Passively reading the literature and published proceedings is no substitute.

It is a good idea to call the program officers of grant programs in the area in which you would like to submit proposals, and let them know you would be willing to help them out by serving as a reviewer. Program officers are always eager to recruit new reviewers, and this is a great way for you to see what is really going on in the field and to network with the people in charge of giving away the money. You will also gain an inside look at what constitutes a fundable (or unfundable) proposal for that agency.

Talk to prominent faculty and researchers at your own and nearby institutions about their perception of the science, where it is going, and what are likely to be the important questions and areas that will surface in the next year. By all means read review articles and the "Perspectives on . . . " books and journals, but remember that publishing something, even electronically, takes time, so whatever appears in print is always bound to be somewhat out of date.

As you work through your survey of the field, you are bound to find, particularly in fast-developing fields with commercial applications (for example, materials science, genetics and proteomics, bioinformatics, computer science, engineering, and minerals exploration), that some of what you want to know cannot be freely discussed or shared by those who know it, for either selfish or legal reasons. This may irk you, but it is one of the realities of contemporary science.

That is why personal conversations with people in confidential settings like Gordon Conferences, which forbid the publication of proceedings from their meetings, are so valuable. In these settings, people are willing to say things on an afternoon stroll or at the social hour about their own and others' work that they would never put in print or an e-mail. Even if all they have to say is, "My advice is not to do that experiment, since two of my postdocs are doing the same thing," that is valuable information. Of course, you are free to ignore it. You are even free to suspect that people who tell you such things may simply want to protect their own turf. Nevertheless, you are better off having had that conversation than proceeding in ignorance. You can always get more perspective on any claims of research activities from others familiar with the person, his or her lab, and the work in question.

You may also wonder if it is wise for you to share your ideas, particularly with someone who may be better equipped than you are to carry them out quickly. That is a chance all scientists take when they communicate and the potential price of the invaluable insight and advice you might gain by striking up a conversation. In my experience, the risk that someone will steal your idea is relatively small because the vast majority of scientists are scrupulous about intellectual property, and those whose history is otherwise are well known to their colleagues and the community. Remember too that you don't have to reveal every detail of your work. You can be collegial and open to discussion, and still play it close to the vest.

When you have completed your fact-finding mission, you will have achieved an independent grasp of your research and field that is deeper and far more profound than you had when you came up with the original idea. You will also have the basis for a better, more sophisticated research proposal.

Choosing Your Work: Being Realistic Without Playing It Too Safe

A common criticism of proposals from people beginning their research career is that the proposed work is too much like their mentor's or is too similar to what the applicant has already done. (What constitutes "too similar" is a judgment that varies from agency to agency. Some, like the National Science Foundation, like to build on strength; others discourage safe, small-scale projects and encourage pioneering thought.) The opposite problem arises when applicants promise too much science for the money requested, for the time that they actually have to do the work, or for the skills they bring to the project. These twinned problems arise when the proposal writer has not done a good job of delimiting the project. Let's address the issues in order.

Staying Too Close to Home

It is, of course, very desirable for one's early research efforts to succeed—that is, to produce clear answers to important scientific questions that result in early publications. But this is also the time when you should be pushing yourself and not be satisfied with incremental extensions of your earlier work, whether it be a doctoral thesis or a note you published ten years ago before you agreed to that temporary administrative job.

When you come up with an independent research idea, let the idea stew for a few days. Discuss it with colleagues, both in and outside the field. Be sure you can formulate your idea and its impact clearly so that a stranger to the field can understand and appreciate what you are going to do, and why. Be receptive to naive questions from people who are not as familiar with your field as you are. They may see more clearly from outside the forest. You will feel much better about your idea and your research plan when you have had a chance to challenge it in this way.

Promising Too Much

An even more common criticism of research proposals from new investigators is that the work they propose cannot possibly be done in the time specified and for the money requested. This is an easy mistake to make, and it speaks well for the scientific imagination of the applicant. Nevertheless, it almost always dooms the proposal.

If you have reaped the intellectual rewards of a thorough review of your field, you may well find that you have more excellent ideas for new work than you can fit into a single proposal. So how could this be a problem? Isn't it good to have lots of interesting ideas? Sure. Eggs and puppies are good things too, but putting too many of them in one basket can lead to a mess.

It is easy to understand why someone starting a research career can make the mistake of promising too much. Neophyte scientists may not yet have an appreciation of the difference between the amount of research they can accomplish in a sixty- or seventy-hour work week as a postdoc and the amount that they can do as a first-year assistant professor with teaching and departmental duties and with inexperienced student collaborators. Add to this the mass of problems generally encountered in starting up a research program, such as buying and setting up equipment, renovating lab space, learning the support ropes of a new setting, and writing proposals and papers. It is likely that, for a while, your productivity will not be what it was before you started your first academic position.

For new and experienced researchers alike, it is always tempting to include ideas in a proposal as they occur during its writing, if for no other reason than to demonstrate to the reviewers and panelists the fertility of one's imagination and the great impact of the idea proposed. But proposals written this way begin to read more like a lifetime career plan than a realistic research program for two or three years. Panels routinely downgrade such proposals because they show that the writer has not carefully thought through the project and the required methodology.

The cure for this problem is simple: in your proposal narrative, carefully distinguish between the finite and realistically planned project for which you are asking support and its possible extensions and elaborations that might take place in future years. Put these latter ideas in a separate, brief, and clearly labeled section at the end of the proposal. That way, you will get credit for all the new ideas, some of which may turn out to be the main point of the work that eventually gets done, without looking like an irresponsible dreamer. We'll come back to this point in Chapter Three.

Expertise: Boon or Bane When Choosing a Research Project?

Staying close to home can take technical as well as intellectual forms. One of the career twists that can occur is for a person to become so expert in a particular technique—for example a new breed of mass spectrometer or a new mathematical model—that the technique, rather than the science and questions that first led the person to master it, begins to dominate the person's scientific career. I refer to this as the "tall ladder" problem (see Exhibit 1.2).

Becoming a "tall ladder" scientist is not necessarily a bad thing. Many scientists have enjoyed distinguished and lucrative careers by acquiring and maintaining the ability to make meticulous, reliable measurements of a certain kind, thereby becoming indispensable to the advance of science because their laboratories ensure that the data on which our ideas are based are reliable.

The opposite approach would be to focus entirely on specific scientific questions, bringing many kinds of experiments to bear on them, and either learning new techniques as needed or collaborating with those who own the "tall ladders" to get the data required. This can have a downside as well. Relative inexperience with an experimental technique can seduce a scientist into oversimplified conclusions, and overreliance on a particular theoretical model can lead to overblown claims about its reality.

There is no right answer or direction to the course of a scientific career. It is a matter of choice and personal intellectual style, but it does have an effect on the kind of research proposals you will be writing and the agencies or programs that

Scientists who become very proficient at a single experimental technique can be like someone who happens to own a very tall ladder that was bought to investigate a leaky roof. Aha! You find the leak, and fix it. Maybe you find there are a few other weak spots in the roof that are more obvious at close range, so you fix those. While you're up there, you clean the dead leaves out of your gutters and maybe touch up the paint. When you finish that, you could get back to other ground-level chores, or you could carry your ladder around the neighborhood fixing other people's roofs and gutters for them.

While such people are valuable citizens of any urban or scientific neighborhood, they run the risk of neglecting their own turf in favor of perfecting an expertise that can take over their lives.

Exhibit 1.2. The Case of the Very Tall Ladder

will fund them. I mention it here because if, for example, you are preparing a proposal to use a scanning tunneling microscope (STM) to measure the rate of alloy corrosion, you owe it to yourself to ponder, at an early stage, this question: Am I proposing this work because I am really curious about the reactivity of metal surfaces and ready to begin years of serious investigation on this important topic, or is it because I have access to an STM?

Aligning the Impact of Your Work with the Aims of the Grant Program

It is crucial when planning a proposal to select a program and funding agency whose missions will be advanced by what you are proposing. If your proposal does not mesh with and specifically discuss what impact the proposed work will have on the mission of the program to which it is submitted, the chances that it will be funded are small.

Sometimes the match between a proposal and the mission and goals of a particular funding agency is obvious, and no explanation is necessary. For example, it is easy to understand why the Juvenile Diabetes Research Foundation would support research related to finding a cure for this disease and not a study of global climate change. However, it is not so obvious what an agency expects when its mission is to support fundamental research that is not directly linked to any particular social or medical problem. Here, the desirable outcomes for the agency may be several and more subtle. When in doubt, call the person in charge of the program.

Usually agencies have several grants programs with distinct program goals, all of which are consonant with the agency's overall mission. For example, an agency may have separate programs for beginning researchers or for primarily undergraduate institutions. It is just as important to respect and adhere to these program goals as it is to understand and respect the overall mission of the agency.

Provided a proposal is in line with the program's and the funding agency's missions, program officers and panelists are more conscious of the *scientific* impact of the proposed work on future research than of any other single factor. The scientific enterprise works when other scientists read with interest the results of your research, challenge it, try related experiments, and in general carry forward the collective dialogue between a research community and the part of nature that it studies.

Also of interest to most review panels is how well applicants themselves understand and describe the intellectual impact of the proposed research. This is one place where all those new ideas you had while writing the proposal can help. Describe the ideas as possible follow-on studies, either by yourself or by interested readers of your results. Even if you do not plan to do all the follow-up studies yourself, you should discuss how your results will feed into your field by opening up new avenues of inquiry.

It is not always easy to place ideas and research to which you have given a large fraction of your short professional career into a larger and longer-term context. Famously, it is hard to see the forest when you are in among the trees—and even more so when what you really care about is the inner bark of the American Elm. So it will not be easy to think realistically about how many other scientists outside your narrow specialty will have their thinking influenced by the results of the research you are proposing. But your proposal will be stronger if you do so. The place to discuss the impact of your proposed work is early in the proposal (in the abstract) and often (in the body of the narrative and in the introduction that precedes it).

Impact may mean something other than "influence on future research in this field." Some agencies look for the project's secondary benefits to specifically targeted populations. Several, for example, want to encourage undergraduates to participate in research. A proposal sent to one of these programs that skimps on or does not discuss at all the role undergraduates will play in the research is doomed.

Other programs may be aimed at giving underserved segments of the population better access to research experience and funds, or at revolutionizing the types of laboratory exercises used in science instruction. Again, it is up to the applicant to demonstrate clearly that the proposed project not only directly addresses agency goals and priorities, but that its impact on those will be substantial.

Every year, agencies receive hundreds of proposals that completely ignore the stated goals of the program to which they are sent. Vow never to do this. Always make it a point to research agencies carefully. Then speak to the program officer of those that look right for your work before you begin to write. Chapter Two goes into the art and skill of matching the agency to your project in much more detail.

Working with a Co-Investigator

I have spoken of you, the writer of a proposal, as if you are working alone except for students and postdoctoral fellows in your research group. That is the case with the majority of scientific research proposals; the author of the proposal intends to be the principal investigator (PI) of the phenomena. However, there may be many reasons for working collaboratively with a co-PI.

A co-PI is a second or even a third scientist who may not be the main instigator of the research but whose expertise is vital to its success and who will participate as a full partner in the intellectual work of planning, interpreting, and revising experiments. Co-PIs may be at the same institution as you or a different one. (In a weaker form of this arrangement, a second scientist assists you without sharing the intellectual direction of the research, usually by providing consultation, specialized techniques, access to materials, or entree to field areas, populations, museum collections, and the like. In this case, you remain the sole PI, and the collaborator either provides the useful consultations gratis or on a fee basis, paid from the grant.)

The advantages of teaming up with a co-PI are clear: you have the basic research idea, but some important aspects of the project are beyond your means or expertise. The co-PI expects to partner in the research planning, use some of the grant, and coauthor the resulting publications but probably would not carry out the research without you. For example, if you are proposing to test a new physico-chemical model of seawater evolution, the participation of an expert mineralogist strengthens your proposal about the conditions under which certain clays react with seawater. Or someone who is a whiz at sorting one subspecies of goldfinch from another may make it possible for you to test a theory of goldfinch migration.

When the co-PI is at a different institution, some agencies do not want to be bothered with splitting a grant between institutions. In this case, one investigator is designated as the main, or "top," PI and her institution the grantee institution. Any funds that are intended for the co-PI flow through a subcontract from the grantee institution to the co-PI's institution.

The wise addition of a co-PI can transform a weak proposal into a very strong one, by answering potential questions about your ability to carry out the research you describe in your proposal. However, there are some special circumstances you need to be aware of and look out for. These have to do primarily with eligibility questions.

First, some grants programs (usually of the "starter grant" type, which I'll discuss in Chapter Two) are intended solely for single investigators. Co-PIs are disallowed in these programs. Second, in any program in which there are restrictive limits on PI or institutional eligibility (for example, grant programs intended for and restricted to nondoctoral institutions), all PIs and institutions usually must meet all of the eligibility criteria. If the co-PI's department awards doctoral degrees, the proposal may become ineligible for support under that program. And finally, some agencies will entertain or carry on their books only one proposal or grant at a time from any individual or department. Presence as a co-PI on one proposal may make an investigator ineligible to submit a second proposal, or to hold a concurrent grant, at those agencies.

Summary

This initial survey of the world of research grants visited each of four levels of review: initial review by the program officer, technical peer review, ranking of proposals (taking into account both of the first levels), and formal approval by a responsible administrator, committee, or board.

There are many more good and worthwhile projects than there is money to fund them; thus it is important to fit your research into the bigger picture by means of a thorough survey of what is already being done, understanding the importance of proper balance between boldness and safety and between under- and oversizing your project relative to the time and resources at hand. The chapter also offered a brief discussion of the relationship between career aims and the type of research one does and the proposals one writes; stressed the central importance of making clear the scientific and agency-priority impact of what it is that you propose to do; and addressed when it is better not to work alone and when bringing in a co-investigator raises new problems.

This chapter has raised more issues than it has settled, and each of the key questions is addressed in a later chapter of this book. I believe that it makes more sense to read about particular issues—for example, what differentiates a cleanly tested hypothesis from a fishing expedition, or when resubmission of a denied proposal is a good strategy and when it's not—after you are acquainted with the context in which those decisions are made.

2

Identifying Funding Sources

This chapter discusses questions you ought to consider before you decide which funding agency, or agencies, you will ask for support.

You may think you already know where to go—usually, one of the big federal agencies like the National Science Foundation (NSF), the National Aeronautics and Space Administration (NASA), or NIH. But keep an open mind. There is rarely a single right agency and almost never a perfect one. Like retail merchants, granting agencies come looking like boutiques, corner groceries, hypermarkets, and everything in between. The concern is not so much that you'll miss the obvious targets, but that you want to be sure not to overlook smaller agencies and special grants programs that might be a promising fit to your project, your institution, and you. Also, grants come from sources other than granting agencies. Lots of corporations, lodges, churches, and individuals will part with small but useful funds under the right circumstances. The key is to remember that you and the funder must share objectives. The chapter also discusses the question of modifying your objectives to match those of the funder, and some special considerations for applicants who are near the beginning of their research careers, as well as for those who are on the faculty of primarily undergraduate institutions.

Y ou may have a strong opinion about which agency you intend to ask for money. Perhaps you are focused on an NSF CAREER grant or a Type G Starter Grant from the American Chemical Society Petroleum Research Fund (ACS PRF). But there may well be other agencies that would be as good or a better bet for your line of research or situation. You won't know unless you investigate alternatives. If you remain convinced that your first idea was the right one, do your best with it. But at least you have accumulated a list of viable backup agencies and grant programs in case your first choice turns you down.

Beginning Your Search

Before the Internet and other public resources, mapping the grantmaking landscape took time, dedication, and often connections that were not necessarily shared democratically across academe. Much of that is gone now that nearly all agencies and foundations have a Web site, where they publish calls for proposals and explain in detail what areas of science, and what sort of scientists, they want to fund. (To get you started, the companion Web site for this book, at www.josseybass.com/go/sciencegrants, has contact information and Internet addresses for a wide spectrum of granting agencies and for pipelines to agencies.) In fact, there is so much funding information on the Internet that the problem is now the opposite of what it used to be, when we scrambled after crumbs of information. Now you have to pan nuggets of useful information out of the rivers of cyberdreck. And just using a search engine is not necessarily the key to success.

Try, for example, putting the words ["Research Grant" + Chemistry] into Google; then try and guess how many of the over 23,000 hits really refer to sources of grants for research in chemistry. Certainly, the NSF home page will be near the front of the list, but so will innumerable pieces from universities about how many of their *chemistry* students are being supported by external *research grants*. Those and others like them will flood out the reference to the Dreyfus Faculty Start-Up Grants, which was listed 190th when I tried this search recently. You need a more sophisticated approach than that.

First, most colleges and universities have a sponsored programs office, which may be anything from part of the job of one subdean or vice president to a great clacking bureaucracy. This office is there to help you get and spend money for your research. Use it! It may be a great resource that can save you a lot of time and energy. These offices often also know of agencies or foundations that have programs to which only investigators from a select group of institutions (possibly including yours) are able to apply, such as the Keck Foundation, the David and Lucile Packard Foundation, or the Howard Hughes Foundation.

Second, there are commercial Web sites that are themselves portals to foundation Web sites—for example,

The Community of Science
   http://www.cos.com/index.shtml

Institute for Scientific Information (ISI)
   http://www.isinet.com/isi/index.html

Sponsored Programs Information Network (SPIN)
   http://spin2000.infoed.org/spin1.stm

Illinois Researcher Information Service (IRIS)
   http://gateway.library.uiuc.edu/iris/

These are four grant databases to which access may be purchased on an institution-wide basis. More than a thousand colleges and universities subscribe to one or more of these services. If your institution is a subscriber (and your dean or sponsored programs office can tell you if it is), your search for appropriate funding agencies should begin here. All of them supply search engines to help you navigate their data.

The job is a little tougher without these services but still tractable. TRAM (http://tram.east.asu.edu) and Grantsnet (http://www.grantsnet.org) focus on research funding in the biomedical sciences; anyone can use them without charge. Columbia University's site (http://cpmcnet.columbia.edu/research/writing.htm) has many links to federal and private foundations that support medically related research, as well as general information on technical writing and proposal writing. Most of the prominent funders—NSF, NIH, ACS PRF, and the Research Corporation, for example—maintain lists of the titles of funded proposals on their Web sites. You can do a keyword search on these to see who got money to do research similar to yours and from which agency.

If none of these tools works for you, a brute force strategy is to search scientific papers for acknowledgments, in connection with a good set of specific terms from your research field.

As an experiment, I pretended to be a geneticist and entered the terms [*acknowledge* + *foundation* + *transgenic*], with the phrase [*support of*] into a Web search engine, and turned up (along with the usual suspects such as NIH and NSF)

research papers citing support from the Organic Farming Research Foundation and the Foundation Fighting Blindness. This is an oddly assorted pair to be sure, but it wouldn't be difficult for anyone looking for a grant in an allied field to narrow things still further.

Then I switched hats, posing as a classical chemist. Substituting *nucleophile* for *transgenic* and adding the word *author* to favor hits on scientific articles produced a paper in which support for a surface science study was provided by no fewer than eight funders: NSF, the Office of Naval Research, the ACS PRF, the Exxon Education Foundation, the ACS Division of Analytical Chemistry, the Society of Analytical Chemists of Pittsburgh, the Alfred P. Sloan Foundation, and the John Simon Guggenheim Memorial Foundation. The first three of those might well have occurred to most chemists, but what about the other five?

In the case of each agency that turned up in my search, it was an easy further step to find their home pages. Each had guides for investigators seeking grants. The point is that the first search yielded only a little over 300 hits and the second only 10, not 23,000.

This might be a good time for you to put this book down and try a fishing expedition with a Web search engine, using terms from your own research specialty. If you get thousands of hits, add more terms, or use an advanced search that eliminates whole classes of hits. (When you find an article or report in your field, search it for "Acknowledge" to find out who supported it.) While you're at it, call your dean or sponsored programs office to see if your institution maintains or subscribes to a database of granting agencies like IRIS, ISI, SPIN, or the Community of Science.

Despite all that's available on the Internet, there is plenty of room for word-of-mouth information. Discuss with colleagues and mentors where they have had success in obtaining research funding. Just remember to treat their input as informational—as in, "Thanks for telling me about The Camille and Henry Dreyfus Foundation; I'd never heard of that one"—not controlling. You are the expert on your research ideas, and you will decide which agencies they fit best with.

Remember that the object of this survey is not to elaborate the obvious but to discover the valuable. Just because I am making something of a point of finding the lesser-known agencies, do not, of course, ignore the big players: NSF, NIH, NASA, the Environmental Protection Agency, the Department of Energy, Office of Naval Research, and others. They control so much money that they deserve a place on your list from the start. But they can be very competitive or have low funding rates; they can be a little hard to work with because they are flooded with proposals; and some are slow to make decisions.

That's why you should also have some of the relatively focused, friendly, and nimble small agencies on your planning list. And don't overlook the government of the state where you work. Many states have science and technology boards that maintain small grants programs. Because only laboratories within that state are eligible for funding, the vast majority of your competitors at federal agencies are eliminated from the start.

Researching Agencies

Let's say that you have come up with a list of a dozen or more agencies that might be good matches to the thrust of your research. How—or how much—do you narrow the list and decide where to put your first efforts?

The key to winning grants is that the objectives of your research fit the objectives of the agency. No worthwhile agency is shy about what those are. Start at the top of your list, and go to each agency's Web site. Read all the documents carefully, and mark a printout with a highlighter pen to emphasize significant information—for example,

- The agency's mission
- Programs of interest
- Eligibility criteria
- Whether the agency supports the purchase of capital equipment
- Any requirements for cost sharing
- Maximum award amounts
- Specified budgetary constraints, such as maximum amounts that can be allocated to faculty salaries or foreign travel

Even in your preliminary survey, you need to be sure that the agency's guidelines and your needs match. If there is an explicit match, you need to send that agency a proposal. If not, don't be discouraged, but move that one down on your list. Always remember that readers tend to see what they're looking for and not necessarily what the writer intended. Don't let your interest in getting a grant lead you to read information into the agency's literature that isn't there.

Agency mission statements are almost always the product of much thought and careful writing. Read them closely. Don't say to yourself, "Well, they *say* they're supporting research in dairy science, but I know they are really broader than that." If you think that's so, you need to hear it directly from one of the program officers before you start to write them a proposal.

At the same time you are checking the mission statement and eligibility criteria for agencies, you also need to get familiar with all of the grants programs that each funder offers. Rarely is there only one per agency. Your objective is to find the best one for you. If you are a beginning researcher, see if the agency has a "starter grants" program in which your competitors are restricted to others who are also beginning a research career, not well-established researchers. If you are from an underserved state, region, institution, gender, ethnic group, or subdiscipline, see if there are special programs for your situation.

Also, be sure that the agency makes grants large enough to support the research you intend to do. Compare the maximum award it gives in each program to the minimum budget you need to do the work. These and dozens of other questions are addressed in the funder's literature and Web site, sometimes in the form of a helpful frequently asked questions (FAQ) page and sometimes scattered through the text. Use your browser's "Find" function to ferret out this information. Be sure to make notes, and write down any questions you might have regarding the program. The notes you take at this stage will prove invaluable when you call the agency and speak to one of its program officers.

A good general rule is that if you are eligible for a restricted grant program at a particular agency—for example, because you are just starting your research career—you should apply first under that program. Funders will be reluctant to spend general monies on you that they could be using to support others who do not qualify for the more targeted program. Time and again, I have heard a panel say, "Great idea; ought to be supported. But they should have come in for a Starter Grant first"—then turn down that great idea in favor of another applicant who would not qualify for the starter grant.

Talking to Agencies

After you have familiarized yourself with the three or four grants programs that look like good matches to your research, call the agency and talk to a program officer. The agency's Web site provides names and contact information. If it's not obvious which name to call, pick one, and expect to be directed to the best match to your question.

Besides being available by telephone, program officers and managers attend many national and regional professional meetings to meet with potential new PIs and inform them about their grant programs. It never hurts to go up and introduce yourself and let them know what you do. Whether you meet them in person or by telephone, program officers will be glad to tell you if your work matches their agency's goals and will encourage you to apply if it does. If it does not, they will often have helpful suggestions about other agencies that might be more appropriate.

Talking to potential applicants is one of the most important parts of an agency program officer's job, so you should let them do it. Their job is to *give away* money, not to keep it. They *want* to generate the largest possible number of competitive proposals for their agency to consider. Commonly, program officers can tell you which of the many criteria listed in the program materials are the ones that count most in ranking proposals. They may also warn you of common pitfalls of proposals that come to their agencies and do not get funded (I will survey those also, in Chapter Five). This is also your chance to ask about the actual evaluation process. How important are technical reviews in the decision-making process? Are the funding recommendations done by a panel or by the program officer? What other information is important in this decision?

Do not postpone or skip this essential step. Some people may find talking to a program officer difficult. If you are shy or modest about your research, your institution, or your abilities, or if you speak English as a second language, overcoming the "chat" barrier may require some effort. But any program officer will be much happier to spend time with you clarifying the agency's priorities, rules, and programs than to spend time later managing a proposal that would have had a much better chance of funding if you had called before sending it in. Also, in the exploratory phases, a telephone call is much more efficient than a several-stage, open-ended exchange of e-mails.

When you call, don't ask what kind of research the agency supports. That's the sort of thing you should learn for yourself before you call. Just describe your research to the program officer and ask if this is the sort of project the agency supports. Do not try to shade or slant your project toward what you think the agency likes to fund. Just be clear, honest, and brief. Save the details for your actual proposal. Too much technical detail just drags out the telephone call and does not address the real question: whether the general idea of the project is a good one for this agency. Confirm the information that you took from the Web site about grant programs, deadlines, and submission rules.

If your research is likely to lead to the development of a patentable invention, ask the program officer what the agency's rules are about patents arising from research that it sponsors. And before you enter into an agreement for your research to be cosponsored by a for-profit corporation that might wish to retain a proprietary interest in the results, be sure that this is compatible with the agency's expectations about prompt publication of your results (see Chapter Seven).

If application forms are not available by direct download from the agency's Web site, this is the time to ask for a set. Always get application materials directly from the agency. Do not use materials from any other source, even your own sponsored programs office. You want complete, up-to-date forms for the proper grant program, and you can be sure of that only if you get them directly from the agency.

Even a brief conversation with a foundation program officer will substantially increase your chances of writing a fundable proposal by clearing up ambiguities in your mind and ensuring that you are applying to the right program for you, your research, and your situation. It will also make sure that you understand the ropes.

Here are three strategic tips to help you in your quest for funding:

- *Ask about the agency's funding rate.* Most program officers will provide current and likely future funding rate information to prospective applicants. Before targeting a program for application, call the appropriate program officer to see if your proposal will have a reasonable chance of success with that program. This information can save you a lot of time and wasted effort and help you target programs and agencies where you have the best chance of funding.

However, be wary. Agencies in a number of areas of the federal government have grant programs with a good funding rate, but there may be little or no new money available in the program. (New money is money that can be used to fund research initiatives by new investigators.) This is because those agencies have made long-term commitments to fund continuing research by specific investigators. It is therefore wise to ask not only what a program's funding rate is, but also if there is funding available for new research initiatives by new investigators and what that funding rate is.

- *Don't fight uphill against a low funding rate.* A solid extramural funding record is commonly a key to success in the academic world, so I suggest that beginning proposal writers avoid grant programs with funding rates of less than 20 percent. This is not to say that you should not apply if you believe the program is tailored to you and what you do. But it is a much better use of time, and less wear and tear on your psyche, to write and submit proposals to programs where your likelihood of funding is reasonable. After you have established yourself professionally and honed your proposal writing skills, most of those other programs will still be there, and your success within them will be dramatically increased.

- *Ask about the funding rate on resubmission.* Some agencies collect statistics on the funding rate of proposal resubmissions, following a previous denial, which can be very useful information. It lets you know whether a denied proposal might be more competitive on its second or even third submission to the same agency. Generally, the encouraging answer is that rewritten and resubmitted proposals do much better than first submissions. (Chapter Six examines this subject in much more detail.) Meanwhile, knowledge is power, so the more information you can gather about a grants program, the easier it will be to get funding.

Funding Sources Beyond Agencies

Nonprofit agencies like churches and civic organizations that may be associated with your institution are also potential funding sources. These entities do not exist to give away money, but there may be circumstances in which they would be pleased to support a student for a summer of research or to fund fieldwork in a specific area. Social and behavioral scientists, in particular, should be aware of this possibility. Beyond that, these sources are so particular to each setting that there is little I can usefully say about them. You, however, can get information about them from your institution's development office—and you should.

Do not overlook the desire of corporations, particularly industries, in and near your institution's home town to be seen as good corporate citizens of that town. Often, this can take the form of the loan or outright donation of surplus, used, or mistakenly purchased scientific equipment, and even the provision of cash grants to your college or university. This is particularly the case when the result is a research student or a piece of gear that can be fitted with a name tag: "XYZ Corporation Research Scholar" or "Gift of The XYZ Corporation."

Work with your institution's development office to make sure that you don't step on the toes of other fund solicitations, but it is often true that corporations will jump at the chance to offload some perfectly usable equipment and even a modest amount of cash (enough to support a summer research student or two, for example) in exchange for a tax write-off and a chance to pose on the local news. And you can leverage the donation at no cost to your institution by counting it as part of a required institutional match for an agency grant.

More fundamentally, if you are working in a field close to possible commercial application, such as engineering, information technology, or earth sciences, you should expect considerable corporate interest in the success of your research. A great deal of fundamental work in these areas is funded by for-profit corporations in exchange for an inside track to the technology that is produced. Remember two things: corporate spending on research fluctuates with business cycles and can leave you high and dry during hard times, and it comes with tight strings attached that may make it difficult for you to consider the fruits of your work your own property. Talk to your academic dean and your sponsored programs office before you explore corporate funding for your research.

Bending to Match the Agency's Priorities

It often happens that there is no agency or grant program whose aims exactly match yours. A potential agency may support a study of crop yields but not if genetic modification is involved. It may support exactly your research but only in a program that demands that you employ undergraduate research students. Or it may support fundamental research that investigates the science behind a new type of laser or sensor, but be forbidden by its charter to support the development of the device itself. Other agencies may support only the development work.

How much should you tweak, modify, bend, or distort your original research idea to fit what an agency considers its priority for funding? The answer lies in the escalating sequence of verbs in that question. No one (agency or PI) objects to a reasonable tweak, but few are willing to distort their research goals just to get (or give) a grant. You and the agency will both eventually be accountable for matching what you say you are going to do with what is actually done. It is up to you to make the judgment as to whether a particular modification of your original idea is a tweak or a distortion and whether it will result in a project you want to do or one you would prefer not to bother with.

This too is a legitimate and necessary area for discussion with a program officer. Sometimes a very minor change in focus or presentation is enough to change an also-ran proposal to a real contender, because that change lets you avoid stepping across an agency guideline. And no matter how carefully a guideline is written in the agency's literature, only a conversation with an experienced program officer can help you see how that guideline governs funding decisions.

In the end, it is because of this question of matching goals, not because you want to get a dozen grants at once, that I encourage you to consider many funders before you decide which one to commit your time and effort to. Like writing itself, the process of choosing an agency is an iterative process. It is a dialogue in the course of which your ideas get clearer, more convincing, and more fundable. The goal is to find agencies and grant programs that are a comfortable fit with your research goals. When you have, you can work on writing your proposal, confident that a good job on that will give you a good chance of funding.

Using Research Grants to Buy Equipment

Grants, as most scientists know, are a good source of support for acquiring expensive instruments and equipment. Indeed, at most institutions, they are nearly the only possibility for acquiring this equipment. Thus, you may wish to select an agency based partly on its encouragement of equipment purchases. This does not mean, however, that it is a good idea to invent a research project just to acquire equipment. Even if reviewers don't spot your real motivation, remember that a research grant is a commitment. Don't commit yourself to research you don't really want to do just for the sake of the funds to buy equipment.

There are currently two good general-purpose NSF programs under which you can make substantial equipment purchases independent of a research project: the Course, Curriculum, and Laboratory Improvement—Adaptation and Implementation (CCLI—A&I) program, housed in the Division of Undergraduate Education of the Directorate for Education and Human Resources; and Major Research Instrumentation (MRI), a program of the Office of Integrative Activities. CCLI assumes that you wish to commit several years to a significant and innovative revision of the way you teach science, particularly the hands-on part of it. The program also expects careful and sophisticated plans to assess the impact of your project and a plan for disseminating the results. The following quotation comes from the program description at the CCLI Web site (http://www.ehr.nsf.gov/ehr/DUE/programs/ccli/): "Projects are expected to result in improved education in [science, technology, engineering and mathematics] at academic institutions through the adaptation and implementation of exemplary materials, laboratory experiences, and educational practices that have been developed and tested at other institutions. Proposers may request funds in any category normally supported by NSF, or funds only to purchase instrumentation." The companion Web site for this book provides an example of a successful CCLI proposal, as well as the program's Web address.

MRI "seeks to improve the quality and expand the scope of research and research training in science and engineering, and to foster the integration of research and education by providing instrumentation for research-intensive learning environments" [http://www.nsf.gov/od/oia/programs/mri/start.htm]. Although successful MRI proposals also must demonstrate that the equipment will substantially improve research and education at your institution, it is probably the first place you need to look if you simply want to acquire modern instrumentation without taking on an elaborate and highly innovative educational project.

In any case, the way to choose among these two and similar programs at other agencies is to call the agencies and talk to the program officers.

Grants for Faculty at Undergraduate Institutions

The strategy of applying for grants under the most restricted program for which you qualify also applies to grant programs reserved for primarily undergraduate institutions (PUIs). These are grants intended to support undergraduate research at four-year and—for some agencies—two-year colleges with no, or no significant, doctoral degree programs. Among the agencies that maintain granting programs restricted to this population are the ACS PRF, the Research Corporation, NIH, and NSF.

The NSF's Research at Undergraduate Institutions (RUI) program currently requires that RUI proposals compete with all other proposals sent to the same directorate in the same grant category. However, the program does adapt review criteria to the institutional setting of the applicant and usually includes undergraduate faculty as peer reviewers among the expert reviewers. Successful RUI, ACS PRF, and Research Corporation proposals from undergraduate colleges are included with the other examples on this book's companion Web site (www.jossey-bass.com/go/sciencegrants).

Again, the temptation may be to go for the bigger grants programs on grounds of either economics (grants restricted to PUIs usually have tighter budget limits) or pride ("Who says I can't run with the big dogs?"). My advice is not to do so unless you feel you must. The budget limits of the unrestricted programs are generally larger because they are intended to support graduate students and postdoctoral fellows, who are much more expensive than undergraduate research assistants. Panels may well be dazzled by your ideas, but they are much more likely to wish that you had applied for one of the grants tailored for your setting than to fund you at the expense of someone who is not eligible for these special category grants.

Do not be fooled by comparable funding rates. Perhaps you know that the funding rate in the general program that is open to everyone is just as high as that in the small-budget PUI-only program. Aren't your chances, then, just as good in either setting? No, because the competitor pools are not the same in the two programs. The faculty mentor who can field a team of year-round, nearly around-the-clock graduate students and postdocs can always generate more results in a given time than someone with a heavy teaching load who has only undergraduate coworkers.

Thus, whatever that funding rate in the open grants program may be, it will be dominated by researchers at doctorate-granting departments. Good stewardship on the part of the agency review panel means that scarce general-program money will be awarded to the doctoral institution most of the time.

Finally, almost every grant program that is directed at PUI science has as a top priority the involvement of undergraduate students in all aspects of research, from the strategic planning, through the grinding-it-out benchwork and the rethinking that comes when some experiment produces unexpected results, to the final write-up, presentation, and publication of results. If your project doesn't fit with that priority, you can expect to receive one of those "we regret to inform you" letters. If you have a track record in involving undergraduates in research, be sure to tell the reviewers and panel about it. This is the primary reason they are giving you funding.

Summary

There is a very wide variety of granting agencies. Practically all of them have sites on the Internet, inviting you to approach them for support. Many of the smaller ones have programs in which the competition is less fierce, and they may be quicker to respond to your approach than the big famous ones. Big agencies, though, have a lot of money, and they can give it away in big chunks.

Your search for research support should start with a list of about a dozen agencies, large and small, that are compatible with the science you want to do and the setting in which you want to do it. When you have spoken personally—by telephone or face-to-face, not just by e-mail—with a program officer at every one of that dozen, you will know which ones and which grant programs are the best fit for your research and the best bet for the time and energy you will spend in writing a winning proposal.

Writing a proposal for a research grant is not the best way to get money to buy capital equipment if that is your real aim. There are specific programs for that; be sure to identify and try them first. You will be more successful there.

As a general strategy, particularly for new researchers, you should take advantage of any grant programs for which you are eligible because of special characteristics of your institution or yourself.

3

Writing Titles, Abstracts, and Narratives

No matter how astutely you have researched and chosen a funding agency, the writing is where your proposal lives or dies. A proposal's main job is not, as it may appear, to get money for your research, but to convince its readers that you have an exciting research project in mind, that you know what it takes to carry it out successfully, and that you are the right person to make it happen. No proposal can do that if the power of your idea is hidden by poor organization and writing. Teaming good ideas with good writing is the purpose of the next three chapters, and we will approach it in three steps.

Chapter Three focuses on the scientific heart of the proposal: the title, abstract, and narrative (which has several working parts). In Chapter Four, we'll discuss the supporting parts of the proposal: budgets, time line, and institutional information. We'll also visit some strategic and ethical considerations in regard to reviewers, parallel proposals, appended text, length limitations, and deadlines. Chapter Five is a review of the ways that proposals can achieve excellence or fall short of it.

In any proposal, the parts that deal with science are relatively easy to list: the title, abstract, and narrative. The structure of the narrative, unlike that of the title and abstract, may be specified in the agency's grant application materials, or it may be left completely up to the writer. Regardless, a good narrative is itself composed of distinct parts that must all work seamlessly together:

- The introduction
- Descriptions of the significance of the proposed research in the context of prior work in the area
- A bibliography that contains references to that prior work and other useful information
- A careful description of the hypotheses to be tested and the methods you are going to use to test them
- The intellectual and other impacts of the research

We will discuss each of these elements in turn in this chapter.

Most agencies have a specific order in which they like to see the various sections of a proposal, but that order is not the same for all agencies. Therefore, this chapter does not focus on one of those forms and walk you through it in that agency's order, because that leaves you with a translation problem if you're applying to one of the dozens of other agencies that do not use that specific format. Rather, I will discuss, in logical order, the key elements that any proposal must have and a strategy on how to gauge approximately how much to write for each.

Remember these principles when formulating your proposal:

- *Address three audiences.* As discussed in Chapter One, your proposal addresses at least three audiences: the program officer, who gauges your proposal's adherence to agency priorities and manages the review process and who may be responsible for recommending its funding or denial; the expert technical reviewers, who focus on the science; and the panel of generalists in your field who combine the first two perspectives.
- *Anticipate readers' questions.* Keep in mind that your job is to describe your proposed work, the subject matter, your credentials, your budget, and all other parts of the proposed work so completely in the space allotted that no unanswered questions will lodge and fester in the minds of any of the audiences reading your proposal. Even secondary questions like, "I wonder what the precision of that technique is?" or "How many samples are they going to include?" or "How many students are going to be supported?" can be damaging to your proposal if your narrative does not address them. In any competition, there will always be proposals that are at least as novel or as interesting as yours *and* that cover all the bases. So try to anticipate ques-

tions that will come up in your readers' minds and answer them in the text before they have time to detract from your well-crafted presentation.

• *Use persuasive rhetoric.* Andrea Halpern (2001) of the Department of Psychology at Bucknell University describes proposals as having three primary functions: (1) *exposition* of the scientific problem and your ideas about it, so that the reader has a clear and accurate idea of what you are proposing; (2) *persuasion* of the reader that the problem and your ideas about it are valid and interesting; and (3) *credentialing* you, your students, and your institution as the right setting to do the work. All parts of your proposal must address these three jobs, to a greater or lesser extent, so they are good ideas to keep in mind as you write. You should ask yourself, "What am I trying to accomplish with this section, this paragraph, and this sentence? What audience am I addressing? Am I leaving reasonable questions unanswered? Does this text do a thorough job of describing the science, persuading the reader, and credentialing me and my setting?"

The Title and Abstract

The title and the abstract carry, word for word, more weight than any other part of the proposal. They are the face that readers see when they pick up the proposal and the source of that all-important first impression. They must therefore be crafted with care to give readers an accurate and complete idea of what the proposal is about. They most definitely are not clean-up tasks to be done when the work of writing the proposal is finished.

Remember that program officers, technical reviewers, and members of a panel must quickly understand your proposal as they read. The title and abstract must therefore speak to each of these constituencies without relying on technical details alone. For most of its readers, your proposal is just one thing in a vast stack of things waiting for them to do. Readers do not want to be mystified or amused, and certainly not misled, while they are trying to figure out what it is that you want to do and why. The object is to inform your reader simply and efficiently.

The Title

To avoid focusing on one particular area of science, I have invented a hypothetical research field: suppose your research is a radiochemical study of social behavior in transgenic planetesimals. Here are four possible titles for your proposal:

"Social Behavior in Transgenic Planetesimals: A Radiochemical Study"

"Planetesimal Herds: Gone Fission"

"Blekinsop's Paradigm and the Sorting of Oblate Planetesimals Subject to Dismutation During the First Five Hours as Measured by N-Pumped Laser γ-Ray Spectroscopy and Post-Mortality Paradigmatic Binning"

"Transgenic Planetesimals"

All of these have something to recommend them: the first is accurate and brief, the second mildly amusing and even briefer. The third is accurate and presumably gives details of interest to technical reviewers (though it is far too long). The last is mercifully brief relative to the third, but it seems to promise that this research will encompass everything knowable about transgenic planetesimals.

The second title may strike its creator as cute, perhaps a breath of fresh air to overworked reviewers. Refrain from going down this path. A "clever" title makes readers work harder than they need to because it doesn't tell them right away what the proposal is about. And while they're trying to figure out your little joke or pun, they will be forming a negative impression of you as someone who maybe is just not serious enough about your work to warrant funding at this time. The keys to a good title are *completeness*, *plain speech*, and *the appropriate amount of detail*.

Only the first title conveys 100 percent useful information to all three of its intended audiences. It gives enough detail for expert technical reviewers to know what lies ahead. And it does not, like the third title, baffle generalists (the program officers and panelists on whom the fate of your proposal depends) who are probably fairly familiar with the subject but may have no deep interest in the gritty details of researching it.

Finally, if possible, avoid titles of the form, "The Effect of Dismutation on Transgenic Planetesimals." They make your research sound like a one-experiment fishing expedition with no theoretical structure, not two or three years worth of professional research. The scope of two to three years of research should be much greater than one cause and one effect.

Exercise 3.1 will give you practice in generating titles that are too short, too clever, too detailed, and just right for the proposed (somewhat more realistic) research projects. (Exercise 3.1a gives suggested responses.) This exercise, which you might want to do with a partner and exchange ideas, will give you practice in

For each of the following proposals, supply a title that is (a) too short, (b) too long, (c) too cute, and (d) just right. See Exercise 3.1a for possible responses.

1. You are proposing research on the role of mutations in the development of the nematode *Caenorhabditis elegans.* You plan to correlate phenotype and genetic makeup using detailed molecular analysis of the worm's genetic material.

2. You are proposing research on the rate and modes of speech modification in young women during adolescent socialization (the "Valley girl" speech syndrome). You plan to study and rationalize the roles of ethnicity, family income, diet, and laryngeal development using widely accepted statistical tools.

3. You are proposing to use molecular dynamics simulations and Lennard-Jones potentials to predict changes in the band gap of semiconductors that have been doped with a variety of transition metals. Then you are going to carry out some experiments using infrared spectroscopy to examine the results of your predictions.

4. Research is proposed on the azimuthal distribution of subsurface water on Mars. Your analysis will combine orbital spectroscopy with Hubble, Viking, and Mars Explorer imagery, radar reflectivity, and planet surface morphology.

5. The proposed research is on biogenic organic compounds in Antarctic ice cores. You plan to correlate compound-specific ^{13}C anomalies in the cores with global sea-level variations and paleogeography to examine the onset of Patagonian forestation.

6. Your research is on forming monolayers of bromine and iodine on graphite surfaces. You plan to examine epitaxy and reconstruction as a function of temperature, precursor, and atomic radius using scanning tunneling microscopy.

Exercise 3.1. Exercises in Writing Titles

what to do and what not to do when creating a title for your own proposals. Pay attention to the process that goes on in your head as you are creating the less-than-perfect titles, so you will be able to recognize the warning signs when you start writing the title of your own proposal. More examples of good titles are found on the companion Web site for this book.

Naturally, there is no single right answer to each task in Exercise 3.1. Following are possible responses that exemplify the advantages and disadvantages of titles that are, respectively, too terse, too detailed, too clever, and, most elusive of all, probably just right for the projects described.

Proposal 1

a. *Too short:* "Nematode Mutations" or "Genetics and Development in *C. elegans*"

b. *Too long:* "Mutation and Nuclear GFP Expression During Embryonic Development in *C. Elegans* at the 4–8–, 20–32–, and 40–48–Cell Stages"

c. *Too cute:* "Teenage Mutant Vinegar Worms"

d. *Just right:* "Mutation and Phenotype in *C. elegans* Development"

Proposal 2

a. *Too short:* "Adolescent Speech"

b. *Too long:* "Correlation of Socioeconomic Factors with Rhythms, Vowel Shifts, and Deglottalization During Socialization of Women from the Age of 11 Years 6 Months to 17 Years 6 Months Observed During Field Surveys in Ventura County, CA"

c. *Too cute:* "She's All, 'C'mon,' and I'm Like, 'Hel–*lo*'"

d. *Just right:* "Speech Modification Among Adolescent Girls: Socioeconomic Factors and Physiology"

Proposal 3

a. *Too short:* "Band Gaps in Semiconductors"

b. *Too long:* "Semi-Empirical Calculation of Band Gap Modification in III-V and II-VI Semiconductors During Group 7, 8, and 10 Element Doping Using Lennard-Jones Potentials and Molecular Dynamics"

c. *Too cute:* "Closing the Band Gap with Lennard and Jones"

d. *Just right:* "Transition Metal Doping and Semiconductor Band Gaps"

Exercise 3.1a. Suggested Responses to Exercise 3.1

Proposal 4

a. *Too short:* "Water on Mars"

b. *Too long:* "Azimuthal Distribution of Subsurface Water on Mars: Orbital Spectroscopy, Imagery, Radar Reflectivity, and Surface Morphology"

c. *Too cute:* "Keep A-Movin', Woola"

d. *Just right:* "Azimuthal Distribution of Martian Subsurface Water"

Proposal 5

a. *Too short:* "Organics in Antarctic Ice Cores"

b. *Too long:* "Compound-Specific ^{13}C Anomalies in Triterpenes of Antarctic Ice Cores: Climate, Patagonian Forestation, and the Global Sea-Level Curve"

c. *Too cute:* "Springtime in Patagonia"

d. *Just right:* "Paleoclimate and Forestation in Patagonia Reflected in Triterpenes from Antarctica"

Proposal 6

a. *Too short:* "Halogen Epitaxy"

b. *Too long:* "A Scanning Tunneling Microscopic Study of Epitaxy and Reconstruction as a Function of Temperature, Precursor, and Atomic Radius of Main-Group Adatoms, Including Bromine and Iodine, on Graphite Surfaces"

c. *Too cute:* "Br. May I Share Your 'Taxy'?"

d. *Just right:* "Epitaxy and Reconstruction in Halogen Monolayers on Graphite"

Exercise 3.1a. (continued)

The Abstract

After the title has given the reader an accurate idea of the topic and scope of the proposal, the job of the abstract (sometimes called the project summary) is to flesh out your idea so that everything scientifically important about your project is revealed to the reader in clear technical language.

Crafting the Abstract

The abstract, a brief section of 200 to 400 words, should address all of the important elements of the narrative that follows. It should include the context and significance of the study you propose, the hypotheses you will be exploring and the means you will use to test them, and the impact (both scientific and otherwise) and implications of the proposed work.

The abstract is, in other words, a road map of the proposal, omitting only such things as the budget, qualifications of the PI and other personnel, the details of methods and materials, and the bibliography. By convention, no literature is cited in the abstract. You will get to that level of detail in the body of the narrative.

Do not minimize the importance of the abstract. Many tentative but important decisions are made about your proposal on the basis of the abstract alone—for example,

- Whether your proposal appropriately addresses agency goals
- To which of several possible program officers and panels or study sections it will be assigned for review and ranking
- Whether it describes cutting-edge science
- Whether its impact is exciting enough for someone to read it with enthusiasm
- Whether your proposal is well focused or flops all over, promising a laundry list of poorly coordinated research ideas
- Even whether the proposal is likely to be recommended for funding if the later details support the first general impression

By the time they have read the abstract, reviewers or panelists may pretty much already have made these decisions subconsciously, even though they will read on through the rest of the text to make sure they have not misjudged the proposal. Thus, the abstract is a crucial part of your proposal, perhaps even the most crucial. It is therefore imperative that your abstract be clear, well written, and completely faithful to what you are proposing.

You may think this last requirement of being completely faithful to your proposal is self-evident, and it should be. Nevertheless, you would be amazed at the number of proposals submitted in which ideas that are introduced in the abstract never appear in the narrative (and vice versa).

It is not hard to understand how this can happen. Probably the abstract was written before the narrative. As the writing progressed, the researcher dropped some ideas and substituted better ones. This almost always results in a better narrative. But too often the writer does not go back and revise the abstract to fit the newly revised narrative.

To avoid this kind of mismatch, write the abstract before you move on to any of the narrative. It will serve as a map of what you intend to say in the proposal and help put your research ideas into a concise, logical form and allow you to get them straight in your own mind. Next, write the narrative, using the abstract as a guide *and* revising the abstract as you introduce changes in the research plan.

When you have finished with the narrative, read the revised abstract you have created. Is it still clear and logical? Is it still a faithful picture of the proposal as a whole? If not, rewrite it from scratch, now using the proposal as a guide.

You might wonder whether you should wait until you're finished with the proposal to write the abstract. In fact, you should always write the abstract first. To keep you on track in the proposal writing process, you always need to be able to summarize your research idea in concise, logical statements, and you need these as a guide as you write the rest of the proposal.

Practice with Abstracts

If you would like to try your hand at writing some abstracts before tackling your own, go back to the six research projects listed in Exercise 3.1. Write a 200-word abstract for any of those that you feel comfortable with.

A useful rule casts the skeleton of an abstract in just three questions and a payoff assertion:

"What's the problem?"

"Why hasn't it been done before?"

"Why can we do it now?"

"The purpose of this research is . . . "

Using this method, you can quickly jot down the most important issues you need to cover in your abstract. From them, you can embellish, modify, and rearrange these thoughts until you have covered all the bases. If you read the funded proposals that are included on the companion Web site for this book, you should be able to find all four of those points stated (at least implicitly) and answered in the abstract or project summary of each.

Another good way to practice abstract writing is to cover up the abstract of a journal article from someone in the field whose writing you (or others you respect) admire. Read the paper, and then write an abstract for it. Compare what you have written to what the author wrote. Carefully study both abstracts, yours and theirs, for strong and weak points. What is it that makes one abstract better than the other? How could one or the other be improved? Could either be shortened? This critical evaluation of your writing, and its comparison with the writing of others who already write well, will be a tremendous help in learning how to write the compelling abstracts.

Something that is always instructive is to go to the Web site of the agency you have targeted and take a look at the collection of abstracts from proposals it has funded. You can readily find examples at the NSF site (http://www.fastlane.nsf.gov/servlet/A6RecentWeeks) or the NIH site (http://commons.cit.nih.gov/crisp3/Crisp_Query.Generate_Screen). Bear in mind when doing this exercise that all the abstracts you see are not necessarily equally good. They do have one thing in common, however: all were part of funded proposals.

The Narrative: The Heart of the Proposal

The narrative is the part of the proposal that carries the real weight. Here is where you make the strong, compelling case that will convince the reviewers and panelists that your work really needs to be funded. A good narrative has some indispensable parts: an introduction; a discussion of prior work and your proposal's significance, along with a bibliography; hypotheses and methods you will use to test them; and a description of other impacts of the research.

Introduction: Engaging Readers

The abstract of your proposal plays one kind of introductory role. It is like the map of a new territory. But it should not be identical to your narrative introduction. In the proposal narrative, you need to provide a different kind of introduction, one that will lead the reader into that new territory.

Experiments have meaning only in the context of a conceptual framework that interprets them. The introduction is therefore the place where you explain the theoretical framework that supports your hypotheses and gives meaning to your experimental tests. This framework allows the reader of your proposal to understand why what you propose to do is important and what impact it will have on future research. You will discuss these impacts later in the narrative, but you will make it much easier for the reader to understand your proposal if you present the conceptual setting of your work early, so the rest of the proposal can be read in that context. Thus, whether you intend to interpret your observations in terms of string theory, early childhood trauma, plate tectonics, or punctuated equilibrium, you must be explicit about the models you will be using.

Your introduction should have another side, too. Every piece of scientific work has some kind of aesthetic appeal that is fundamentally nonrational (Blackburn, 1971). It may be the joy of being out in the field with the rocks, the animals, or the crops; it may be the quiet planning and execution in the laboratory or the thrill of admiration at how nature takes care of its business by its own rules, which you are privileged to understand and share. It is the motivation that makes you want to do a particular piece of research. Whatever your field of science, you do what you do because you get pleasure or personal satisfaction from it. Thus, you are motivated by something outside the world of measurable quantities, hypotheses, deliverables, control groups, budgets, or grants.

There is a place in your proposal for this aesthetic motivator, and that is at the beginning of your narrative, in the introduction. This is where you can make that point (though not, of course, in these words): "Here, folks, are the weird fish, the cool patterns, the distant worlds, that I plan to investigate." Invite the readers into your laboratory, down the barrel of your microscope, to your outcrop, your subject population, or your rat maze. Use a visually appealing picture or graph that illustrates your research. Sharing some of your excitement about the field helps prepare readers for the hard work of careful analysis of your proposal. Then get right down to business by telling them who else has already walked in this territory or down these paths.

Prior Work and Bibliography

Your goal here is to get readers to understand the problem as you understand it. Secondary jobs are (1) persuading readers that what you are proposing is a vibrant and worthwhile area of research and (2) credentialing yourself as someone who has done your homework. The anchor of any prior work section is the bibliography, in which you give credit to all those giants on whose shoulders you are preparing to stand, setting the work you want to do into a recent historical and intellectual context.

Proposals are about the here and now, so your discussion of the literature should be up-to-date. Although citing original or pioneering literature is justified in some cases, by far the vast majority of your citations should be recent work, including, but not dominated by, your own contributions. Your preliminary work surveying the state of your field, described in Chapter One, has prepared you to write this section. You are persuading the reader with your concise prose that the field of work you are about to embark on is intriguing, multifaceted, and active and that you have a significant contribution to make to it.

Here are a few tips to make your supporting references really count:

- If the grant application does not include instructions as to the format of bibliographic entries, call the agency and ask. Most agencies use the same format as the leading journals in your field. Unless otherwise instructed not to, always provide full citations. A proposal's bibliography provides a crucial resource to reviewers and program officers, many of whom will use it to help them assign appropriate reviewers or determine possible reviewer conflicts of interest. For that, they need to see the full citation, including title and the names of all authors (do not use "et al"). Provide this information.

- Don't include every possible literature citation in the bibliography. Except in unusual circumstances, the length of your bibliography, depending on your field, is usually adequate at fifty to one hundred (or so) citations. This range allows you to do justice to the important work in your field while showing readers that you can distinguish important literature from what is less important. A very short bibliography implies either that you're working in a very minor field of science or that you're not very well acquainted with what's happening in it. A lean bibliography also risks slighting important workers in the field, some of whom may review your proposal. More than two hundred citations looks like tedious overachieving or like you can't tell the wheat in your field from the chaff. It also raises the legitimate suspicion that you haven't read everything you cite.

- As a corollary to the previous point, include only work that you have read and thoroughly understood. If you do not, Murphy's law will ensure that you will end up misquoting and misusing other scientists' ideas. These very people will no doubt be assigned to review your proposal or, worse yet, be on your panel.

- Don't be modest about citing and critically discussing your own previous work. This is where you get to demonstrate that the work you propose builds on strengths that you already possess and work you have done and transcends the limitations of your previous work.

- Be sure that the literature you cite is a faithful mirror of your proposal narrative. Don't cite articles that you do not refer to in the text, and be sure that you cite all that are. Verify the bibliographic information yourself; don't copy it from other sources. And by all means, don't import other people's bibliographies wholesale into yours. This labor-saving maneuver can lead you to include articles that have nothing to do with the proposal and will propagate typos and other errors, making you appear to be the kind of lazybones no one will want to support with the precious few funds that are available. Do not imagine that no one will ever notice additional or omitted citations. It is common for even a single missing or superfluous reference to be pointed out by more than one reviewer.

- Be sure that very recent work is well represented in your literature survey and your bibliography. Otherwise, readers may conclude that you're working in a dead or dying field. An exception might be the case in which you intend to shed new light on an old and unsolved, but still significant, problem. In that case, your job is to convince readers that the problem is still interesting and that your work brings a completely new approach to it.

- The size limits of proposals for most major funding agencies do not include a space limit on the proposal bibliography. Therefore, do not try to cram your bibliography into the smallest space possible, using small fonts, thin margins, and eliminating spaces between references, or by chaining references into long strings. This formatting makes it difficult for program officers, reviewers, and panelists to discern whether you have cited the relevant literature and the key players in your field.

- A corollary to the previous point is not to include substantial technical or editorial remarks in your bibliography, because you will probably be seen as trying to evade the narrative length limits. Besides, if these are important points you're making, your reader is likely to miss them if they're tucked into the bibliography.

Research Impact and Significance

The purpose of research is not just to find things out—the shapes of gold nanoparticles capped with different kinds of dendrimers, the structure of DNA, the solubility product of lead sulfide, the thermal history of the Texas Permian Basin, the ontogeny of the panda's "thumb," or the eccentricity of Pluto's orbit. Interesting as these factual questions are and as satisfying as it may be to know their answers, their real life lies in their impact and implications in relation to broader insights and further research. For example, why is lead sulfide so much less water soluble than calcium sulfide? What molecular-level factors govern its solubility, and under what circumstances do lead sulfide ores form? What is the cause of the maturation state of hydrocarbon deposits in the Permian Basin, and what other basins worldwide might be similarly oil rich? What does the eccentricity of Pluto's orbit say about the history and likely future of the solar system?

Agencies prefer to fund research that not only answers questions but also leads to new questions and other possible avenues of research. Very few are tempted by proposals aimed at generating lists of uninterpreted data (such as tables of solubilities, lists of species, or surveys of ethnicity) or simple characterizations of materials. Even tightly mission-oriented agencies like those targeting specific diseases are interested in more than just the technical elucidation of the disease. They want to fund work that will help to translate theoretical insights to clinical applications. Your proposal is unlikely to succeed unless you persuade the reader of the impact and wider implications of your work.

You have presented the theoretical context of your proposal in the introduction. Further discussion of the significance of the research you propose should be woven throughout your narrative. But lest the reader miss it, there should also be at least a half-page, so labeled, that discusses nothing else. You can develop this section from the list of possible extensions of your core project that I discussed in Chapter One.

Exhibit 3.1 shows how one major funding agency, NIH, values the significance factor. If you are writing a proposal to any of NIH's institutes, you must consider and address these criteria throughout your proposal. Regardless of which agency you are addressing, the five criteria set out in the exhibit should always be in your mind and addressed in your proposal.

Think about the research projects you want to pursue, and make up a preliminary list of the impacts the work will have on both the topic at hand and the larger science. Do not forget to list intangible impacts also, such as the training of undergraduates, mentoring and education of underserved populations in the sciences, or the use of a requested piece of equipment in laboratory classes. Go to the examples of funded proposals at this book's Web site (www.josseybass. com/go/sciencegrants). Locate and read all the places where successful proposals mention the significance of what they want to do. Additional significance statements can also be found in nearly all of the abstracts of funded proposals posted on the Web by NSF and other funders.

Hypotheses and Methods

This is where you focus on the guts of the project: what you're going to prove and how you're going to prove it. Make sure that this section takes up the lion's share (80 percent or so) of the proposal narrative. The parts I have discussed so far mainly cover the theoretical context of your research and what others have done in that area. Reviewers and panelists really want to know what *you* are going to do. It is this section of the proposal that, in the end, will get you your support. I have seen proposal after proposal passed over because of a narrative composed of 75 percent introductory and historical material and 25 percent everything else. Don't minimize the amount of information you supply here about what you plan to accomplish and how you are going to do it. You can't count on funding based only on your knowledge of the previous literature or how your work fits into the bigger picture.

A good example of the importance of the "significance" factor is its placement at the head of the criteria that the National Institutes of Health uses in its review of research proposals:

Review Criteria for and Rating of
Unsolicited Research Grant and Other Applications

1. *Significance:* Does this study address an important problem? If the aims of the application are achieved, how will scientific knowledge be advanced? What will be the effect of these studies on the concepts or methods that drive this field?

2. *Approach:* Are the conceptual framework, design, methods, and analyses adequately developed, well-integrated, and appropriate to the aims of the project? Does the applicant acknowledge potential problem areas and consider alternative tactics?

3. *Innovation:* Does the project employ novel concepts, approaches or methods? Are the aims original and innovative? Does the project challenge existing paradigms or develop new methodologies or technologies?

4. *Investigator:* Is the investigator appropriately trained and well suited to carry out this work? Is the work proposed appropriate to the experience level of the principal investigator and other researchers (if any)?

5. *Environment:* Does the scientific environment in which the work will be done contribute to the probability of success? Do the proposed experiments take advantage of unique features of the scientific environment or employ useful collaborative arrangements? Is there evidence of institutional support?

Exhibit 3.1. How the National Institutes of Health Views "Significance"
Source: http://www.nih.gov/news/NIH-Record/07_29_97/story04.htm.

Testable Hypotheses

Many proposals limp into the outer darkness with the epitaph: "No testable hypothesis. It's just a fishing expedition."

Fishing expeditions are fine, if you're writing *A River Runs Through It*. And they play a vital role in science, but rarely a fundable role. Scientific fishing expeditions go by the aliases, "Mix them and see if they react," or "I bet we'll find ethnic correlations," or "Wow, I wonder if that funny peak is a magnetic monopole?" or even, "I want to synthesize and characterize this interesting material." These expeditions can also be as vague as, "What's going on here?" when what is really meant is, "Why does the spectrum of substance A in solvent B have that funny bump at 450 nanometers?" which itself is only a partly defined question.

Of course, most worthwhile science starts out with ill-defined curiosity. These are the questions that you might well turn up in the course of a scientific fishing expedition in your lab. But they are usually not questions that agencies fund, because the question has not been formulated in a way that is *guaranteed* to give a significant scientific result. Panels care about significant results.

This can be baffling and irksome. After all, don't colors and patterns, striking phenomena, the quirks and glories of nature count for something? Of course they do. You've already covered that in the introductory part of the narrative. But unless research has been carefully formulated to produce results, even meaningful negative results, an agency is taking a big gamble in funding it at the expense of less intriguing but better-planned research.

If there was an unlimited amount of money in the world to support research, funding ill-focused curiosity wouldn't be such a problem. But remember that there is not enough money to support even half of the "Very Good" research projects. It would be doubtful stewardship on the agency's part to gamble its precious financial resources on what might turn out to bring home a mere basket of shiny, gasping facts—a fishing expedition.

You avoid fishing expeditions by addressing yes-or-no questions to nature that your protocols can answer. It is not "What happens when I mix these substances?" but, "Does the mixing of A with B produce phenomena [spectral shifts, heat flow, phase changes] that will register unambiguously, reproducibly, and quantitatively on the apparatus I'm going to use?" And it is not, "What is the C-13 content of stalactites?" but "Does the δC-13 of stalactites correlate to a defined level with global mean temperature during times of known global warming, and with what confidence can we use them to calculate mean temperatures at other times?"

Notice that for both of these cases, the better-defined question is also much narrower. That means that you are free to add other well-formulated questions, and your proposal should include as many of those as you can keep under one general topic and as you can answer over the duration of the grant. When you've answered enough of those, you may find yourself invited to present broadly titled talks like "A-B Chemistry" or "What Speleothems Can Tell Us." Just don't expect to have proposals funded when they are as nebulous as that.

You also need to let the reader know how you intend to answer these crucial questions. What kind of samples are you going to use? How will you get them? What kind of response are you going to look for? What kind of precision will you need to tell whether your results are significant? What kind of analysis are you going to use? A brief statement of the things you are going to seek and how you are going to know them will let readers see that you will be able to come to some conclusion. And it will serve as an introduction to the highly detailed information you present in the methods section.

As in the title and abstract sections, the best kind of practice is to try to write meaningful (and less meaningful) versions of what you want to do. To improve your ability to write a fundable hypothesis or a series of them, and the tests you will be using to gauge the results you will obtain, try formulating your own principal research interest first as a fishing expedition question and then as a series of testable hypotheses. You will find hypothetical models in Exercise 3.1 and real ones in the proposals presented on this book's companion Web site.

Methods

This is where you tell readers exactly what you're looking for, how you are looking for it, and where and how long you're going to look. If it's carbon-13 in stalactites, explain your choice of samples, your sampling protocols, your mass spectrometer (or the one you're going to buy time on), and what you are going to use as standards. If it's quantitative spectroscopy, discuss the resolution of your instrument, detection limits and precision, method blanks, and so forth.

Don't give details of well-known techniques if everyone in your field recognizes them by name. But if you are going to modify the technique, explain how, why, and what improvement will result. If the modifications are significant, you should have previous publications that establish the validity of the resulting data. Refer to them here.

This is the place to establish your credentials in using any nonroutine laboratory or field techniques. Here also is where you document that you have access to any special collections, populations, or field areas that a reader might not expect to be routinely available. Remember in this last case that if you plan to use any special data sets that you do not already have in hand, you will need to produce letters of collaboration from the owners of the data or materials that you say you are going to use.

The methods section is the place to explain why you are using the experimental techniques you have chosen. The fact that you have access to a particular instrument does not of itself support its use for the particular problem you're pursuing (remember the case of the very tall ladder from Chapter One). Each experimental technique you use must be at least completely adequate to the tasks you will be carrying out with it. Otherwise, it will be too easy for a reviewer or a panelist to question why you're not using a different and better one.

You should describe the typical experimental results you expect and how you will interpret the data—or their absence, in the case of negative results. If you are looking for a signal, your results may need to be stated in terms of the detection limit for that signal, so that even a null result can produce positive statements of the form, "Using the [described technique], we will be able to show that transgenic planetesimals will emit no more than [detection limit] watts per square meter per second." Such results, depending on the state of knowledge in your field, may well be of fundamental, and therefore fundable, importance. Good examples might be Michelson's search for the ether drift or the many attempts to detect charge, parity, and time reversal (CPT) symmetry violations in physics or to put an upper limit on any number of psychometric correlations.

Finally, don't forget safety, management of environmental hazards, and human or animal experimentation guidelines. The NIH application includes specific mention of these, but others do not. If you leave them out, a legitimate conclusion would be that you didn't think about them and that you may be a danger to yourself, your students, or your subjects. Exercise 3.2 will give you practice in addressing these issues, and Exercise 3.2a provides sample responses to the issues raised.

Other Impacts of the Research

I have already cautioned you about not overloading your proposal by promising more work than you can carry out within the time and budget scope of your project. However, you do not want to give the impression that your work will be without further scientific impact. In a separate section, you should explore the work your research may give birth to in the future, whether it is done by you or by others who have become inspired by reading your initial results. A reasonable (not inflated) list of anticipated further work demonstrates that you have thought through where your research fits in the larger structure of your field and that the work you propose to do will have impact beyond itself.

Also, be sure to address all of the programmatic and ancillary goals that the agency lists as important in its description of the grants program to which you are applying. If your proposal is written to a program designed to promote undergraduate research participation, spell out carefully and explicitly how those students will share in the research, what they will be doing, and what those tasks will teach them about research in general and about your field of research in particular.

Other Considerations

No agency will leave you wondering about its programmatic goals. All articulate them carefully in their literature and Web site. You should address them equally carefully. It is discouraging for a program officer to see excellent research ideas turned down for funding because the investigator just assumed that it was obvious how the work applied to the goals of the program to which it was submitted. Assume nothing. Be explicit in your description of the fit of your work to the agency's goals.

Exhibit 3.2 (p. 58) provides examples of two agencies that explain their criteria. One of those agencies, NSF, goes on to spell out clearly how program officers will apply those nontechnical criteria in recommending funding. You can be confident that a scientifically sound proposal that addresses them head on, point for point, will be very competitive for funding. One that ignores them has virtually no chance.

Following are some examples of research proposals in which reviewers might raise questions on safety, subject treatment, and environmental issues. For each proposal, try to identify what those issues might be, then jot down some notes relating to how you could neutralize or mitigate any possible concerns that reviewers might have. See Exercise 3.2a for some possible responses.

1. Preparation and characterization of organobismuth complexes of the Group 12 metals

2. A field study of structural geology in the Magdalena Valley complex of Colombia, including territory that is subject to guerrilla incursions

3. Impact on standardized test scores of uniform dress codes in urban middle schools

4. Healing of fractures in the long bones of dogs

5. Sensitivity of marine mammals to ultralow sound frequencies

6. Naive-subject response to apparent ethnic profiling in airports

Exercise 3.2. Addressing Questions of Safety and Ethical Issues in Research

The Reality Check

The abstract and narrative are the heart of your proposal. When you are satisfied that you've done your best on them, you should get other scientists' opinions of them. Ask them to check not just for the science, but for clarity, impact, and conviction (I have more to say about conviction in Chapter Five). Although you do want a careful technical check on your science, you should include a reader who is a bit outside your field and can understand the main thrust of the ideas without getting bogged down in technicalities. This person, who will be a stand-in for those generalist panelists and program officers your proposal must impress, will be the best judge of whether your project is something that must be done or is just an acceptable piece of incremental science.

Following are the issues that occurred to me in posing the problems set out in Exercise 3.2. Investigators with more experience in each field will probably think of others.

1. *Organobismuth complexes of Group 12 elements:* The first and primary concern here would be toxicity of both the metals and the ligands (and, indeed, of the adducts). Exposure of students and others in the laboratory to vapors and spills must be eliminated, and any disposal of waste must be into a sealed and traceable waste or recovery stream. Secondary concerns would be flammability or other uncontrolled reactions with ambient air or water vapor.

2. *Colombian geology:* Personal safety of the investigator and students would be the primary concern. Secondary concerns would be trespass on property and prior access by commercial concerns, such as petroleum companies.

3. *Dress codes and standardized test scores:* How will the studies be structured to isolate the dress code variable? Will students be informed of the nature of the study, and will they and their parents be given opportunities to grant or withhold consent? Will participants in the study be compensated? What effect will the study itself have on standardized test scores, and what use will be made of those scores in the future? Have all federal guidelines for human subject treatment been followed?

Exercise 3.2a. Suggested Responses to Exercise 3.2

Give these readers plenty of time, with a realistic deadline, to read your embryonic proposal and get back to you with criticisms. Offer incentives (like reporting to you over a good meal and even reading *their* proposals) that will increase their motivation to read carefully. And then listen carefully, nodding, making notes, looking appreciative, saying "Mm-hmm," and not stopping them to refute or explain what may strike you as trivial or wrong-headed points. (You can certainly show your readers how wrong they are, but winning this battle more than once or twice will mean losing the sympathy of your friendly critics.)

This is your first practice in getting reviews on this proposal and is different from the real thing only in that, at this stage, it is not too late to do something about the criticisms. I address reading and responding to reviews more fully in

4. *Healing of fractures:* How will the subjects be acquired? Will the study use domestic dogs with accidental fractures brought to a veterinarian for treatment or laboratory dogs? If the latter, will healthy dogs be traumatized by the administration of a controlled bone fracture? If the former, how will natural variables in the fracture itself and its surrounding trauma be taken into account? How will healing be assessed? Have all federal guidelines for animal research been followed?

5. *Sound sensitivity of marine mammals:* How will effects be observed? Sound travels surprising distances in the ocean; what radius around the sound source will be included in the study? Will behavioral as well as physiological impacts be included? How long in time will the study last? What will the source of sound be, and what effects have previously been demonstrated in laboratory settings? How will ambient noise variations, such as propeller noise in shipping lanes, be taken into account?

6. *Ethnic profiling:* Since "naive" subjects are posited, how will informed consent be obtained? Will subjects be deluded or tricked during the course of the interaction, even if the deception is revealed later? What permissions and cooperation will be needed from the Transportation Safety Administration? What external characteristics will be used as the basis for profiling? Will foreign as well as U.S. citizens be subjects? Will subjects be compensated and have the opportunity to opt out of the study? Have all federal guidelines for human-subject experimentation been followed?

Exercise 3.2a. (continued)

Chapter Six. You are invited to skip up to the section called "The Four R's of Responding to a Denial" and apply the advice there to this preliminary version of a review.

Summary

Complete proposals consist of a scientific core: the title, abstract, and narrative and other necessary information. The title and abstract are, word for word, the most crucial parts of your proposal because your reviewers will base their important preliminary impressions on them. They must be concise, accurate maps of the proposal as a whole.

If your proposal concerns agriculture, for example, you will find on the Web site of the Organic Farming Research Foundation (http://www.ofrf.org/) a statement of its criteria for spending money on your research. (One is that all supported projects have to be carried out on farms.) At the other end of the size spectrum, the National Science Foundation takes pains in the Review Criteria section (http://www.nsf.gov/pubs/2002/nsf022/nsf0202_3.html) of its Grant Proposal Guide (http://www.nsf.gov/pubs/2002/nsf022/nsf0202_1.html) to spell out exactly what it is looking for in the way of nontechnical impacts:

How well does the activity advance discovery and understanding while promoting teaching, training, and learning? How well does the proposed activity broaden the participation of underrepresented groups (for example, gender, ethnicity, disability, geographic, etc.)? To what extent will it enhance the infrastructure for research and education, such as facilities, instrumentation, networks, and partnerships? Will the results be disseminated broadly to enhance scientific and technological understanding? What may be the benefits of the proposed activity to society?

Exhibit 3.2. Examples of Agency Programmatic Goals
Source: National Science Foundation, *Grant Proposal Guide* (NSF 02–2), Dec. 2001.

The narrative consists of an introduction, including a survey of prior work that is supported by a good bibliography. The research you propose should have a clear theoretical framework and should promise a significant impact on your field. Your research needs to pose well-formulated hypotheses that are tested by well-characterized methods already accepted in the field or supported by previous use in your laboratory. You also must consider safety, environmental impact, and the ethics of animal or human experimentation.

The successful proposal also explicitly addresses agency priorities, which may include nontechnical impacts such as those on target populations, science education, and infrastructure improvement.

Ask a colleague to read and criticize your narrative while you still have time to benefit from an outside review.

4

Preparing Budgets and Supporting Information

This chapter addresses the businesslike infrastructure that accompanies the scientific part of your proposal: how to draft a good budget and how to answer the numerous additional questions that agencies address in their application forms. This material is not as much fun as thinking about science, but it has a crucial role to play in supporting your proposal. Even the very best science goes unfunded when it is not set in a businesslike context by the rest of the proposal.

This chapter also discusses some of the ethical and strategic considerations that arise in the application process: whom to suggest as technical reviewers, what to do about proposals you've submitted to other agencies, and how to manage narrative length limitations and submission deadlines.

Budgets

Almost no one will give you money without asking you how you are going to spend it. In most cases, though not all, agencies expect to see a year-by-year budget of what you propose to spend in pursuit of your science.

Budget forms, budget rules, and limits to overall dollar requests and requests for individual lines within budgets vary greatly from one agency and from one grant program to another.

Exhibit 4.1, from a proposal I once wrote, is a simple example. The budget explanation is provided in Exhibit 4.3 and the complete proposal is shown on this book's companion Web site (www.josseybass.com/go/sciencegrants). There are longer and more complex budgets in the examples of proposals on this book's Web site, approved and funded by a variety of public and private agencies.

	For the Periods (Each period should end on August 31 and be of at least twelve months duration.)	
	Aug. 1, 1988 to Aug. 31, 1989	Sept. 1, 1989 to Aug. 31, 1990
1. Summer Salary: Principal Investigator (Maximum: $4,000 per year including benefits)	$ 4,000	$ 4,000
2. Stipends: Undergraduate Student(s)	1,000*	1,000*
3. Expendable Supplies and/or Services	700	500
4. Capital Equipment (Specify item and any match in narrative)	7,500	-0-
5. Travel (Maximum: $2,000 per year)	-0-	-0-
6. Field Work	-0-	-0-
7. Departmental Allocation (U.S. only; not to exceed $500/yr.)	500	500
ANNUAL TOTALS	$ 13,700	$ 6,000
TOTAL REQUESTED		$ 19,700

*To be supplemented from institutional funds. See budget explanation.

Exhibit 4.1. Budget for an American Chemical Society Petroleum Research Fund Type B Grant

We begin with a collection of commonsense guidelines that submitters of proposals nevertheless violate regularly:

- Don't exceed any specific category limitations imposed by the agency.
- Don't ask an agency to fund any item or type of work that it does not allow.

- Don't ask for more (or less) money than you need to carry out the proposed work.

Beyond these, there are some general principles that will help you prepare a budget that complements the science in your proposal.

In a sense, year-by-year budgets are really no more than good-faith estimates. No one can describe to the dollar how much test tubes or special mice will cost three years hence. Predictions about the rate of spending for ancillary personnel, travel costs, expendable supplies, and even capital equipment are understood to be the investigator's best guess and not an ironclad guarantee. (Chapter Seven provides a discussion of revising these predictions in the light of the progress of your research after it is funded.)

Nevertheless, you do have to put down something. My best advice to you is to read the agency budget guidelines carefully. Call the program officer if something is not perfectly clear. (Program officers probably receive more telephone calls about budgets and changes to them than all other subjects combined—except, perhaps, "Why didn't I get funded?")

Following are some of the budget categories you will have to think about and to research carefully in the agency guidelines. For each, I offer what I have observed to be a good general strategy used by funded proposals. You should also take advantage of the expertise of your sponsored programs office and your colleagues who have been successful in their pursuit of research dollars as you work through your budget.

The Bottom Line: How Much Should You Ask For?

Knowing how much money to ask for can be a fine art, especially for agencies that do not have specified budgetary limits. You should do your best to put aside dreams of large, round numbers ("Wow! We could get $150,000 a year!"). Build your budget up from a zero base. In the best case, this will lead you to asking for at least as much as you need, and not grossly more.

Reviewers and panelists also run research programs, so they know what the real costs are for most things that you want to do. A proposal that comes in "fat" or with a budget containing costs that are out of line with what most people know to be the going rate will be downgraded. A similar fate awaits proposals with budgets that come in "too light." You are never doing yourself a favor by trying to promise something for nothing. Credibility is the gold standard in the world of science, and you need to show both the panel and the reviewers that you are fiscally serious and realistic about your work.

Part of selecting the agency to which to apply is the question of whether the grants available from that organization are commensurate with the cost (and scope) of what you are proposing. You certainly do not want to pitch an ambitious multi-year research program to a foundation that has an award limit of $20,000 and a grant term of one year.

Sometimes maximum awards and the number of years over which the money can be spent are specifically indicated in a program's application materials or on the agency's Web site. For example, the Web site for ACF PRS (www.chemistry.org/prf/) answers the question, "What is the maximum budget, and the maximum PI stipend, currently allowed by ACS PRF Type AC grants?" in this way: "$120,000 for 3 years; $7,500 per grant year" as of this writing. But this is not the case for most large federal funding agencies like NSF, NIH, the Department of Energy, NASA, or the Department of Defense.

One good way of gauging an acceptable funding request is to look at a list of previously funded proposals on the program Web site to see what the range of awards and the average amount seem to be. This will be your guide to deciding whether to select this agency for your proposal and determining how ambitious you can make your project.

Even if your proposal is funded, you may not always get the full amount you ask for. So as you prepare your proposal, you should be thinking of how to scale back your program a bit if you are asked to shorten your project by a year or to trim 20 percent off the total. Remember that you may be asked to trim your budget, but you will almost never be asked to increase it, so be sure that you ask for what you really need to carry out your work.

Also keep in mind that the shorter the time period or smaller the amount of money you request, the more proposals you are going to have to write. Proposal writing can chew up a big chunk of your time, so don't be afraid to be bold and try for longer and more ambitious projects.

Before submitting your proposal, go over your budget with your sponsored programs office, if your institution has one. The staff members there have a great deal more experience in these matters than you do and can be very helpful.

Never sign a proposal with a budget that you don't thoroughly understand. This is a legal document, and you will be held accountable for what you spend and the categories into which those expenses fall.

Major Budget Categories

You've made sure that the agency will support a grant large enough to account for your total budget. Now let's look at the major budget categories.

Faculty and Technician Salaries

It is gratifying that agencies recognize that if you were not doing this research, you might have to clerk at Burger Bob's for a summer salary, so many of them are willing to pay you to do what you would probably do for free in any case. All agencies have clear maximum-salary guidelines, expressed either in dollars or as a fraction of your academic year salary roughly proportional to the time you propose to spend on the research. Abide by them. (All agencies also have stories about investigators who ignored or subverted salary rules. Don't be an anecdote.) Also, be sure that you are clear on whether these maxima apply to *each* of several co-PIs or to all of them combined. Different agencies do it in different ways.

Some agencies also recognize that the effective running of large laboratories or the adequate maintenance of specialized equipment needs the long-term care and dependability of a committed technician. For these, claim only the actual percentage of time that the technician will spend on the work outlined in your proposal. For example, if you have a technician who runs your mass spectrometer but you have three different projects that are collecting data on it, do not charge a full year of the technician to one grant. That's asking one agency to pay for work done for a different agency, possibly including work that is contrary to the first agency's priorities. You need to make a reasonable estimate of the amount of time (generally calculated as dedicated months of service or percentage of time) that the technician will be working on this particular project or training the students who will carry out the analyses.

Some agencies prefer to reserve their funds for the PI, students, and postdoctoral fellows and do not fund in-house technical support. If it is not allowed, do not request it. If you have questions as to whether technician salaries are an allowable expense, read the program's Web page. If the information is still not clear, call the program officer and find out. It is frustrating and unnecessary to have your proposal turned down because you asked for something that is impossible for the agency to support.

Do not assume that all agencies will allow you to pay salaries or stipends of individuals who are not faculty or students at your home institution. This means, for example, salaries or stipends of visiting scholars or faculty members who may be doing a sabbatical with you or perhaps a former student who is between an undergraduate degree and coming to your lab as a graduate student in the fall. Invariably, you need to check first in the agency rules or with the program officer to make sure such allocations are allowed.

I have more to say about salaries and budget management in Chapter Seven. But you can manage effectively only what has been built into your budget in the first place.

Student and Postdoctoral Fellow Salaries

The stipends paid to graduate students and postdoctoral fellows may vary from one department to another of the same institution, and they do vary by more than a factor of two from one part of the United States to another. The discrepancies are even wider between the United States and other countries. Thus, neither agency guidelines nor this book can be very helpful in budgeting them. You should be guided by your department chair or your sponsored programs office in setting the stipend level of students and postdocs. If the official channels have no helpful suggestions, check what is being paid at comparable institutions in your region.

From the student's point of view, working with you should be an education and a privilege, not a blood sacrifice. Don't make your students choose between working with you or taking a summer job or teaching a section of a regular course for twice the money. You and the agency both want as much of a student's attention as possible to be directed toward the research project.

When funding graduate students, do not forget to find out whether tuition is an allowable expense. Agency policies vary on this question. Read the guidelines, and talk to the program officer.

Fringe Benefits

Some agencies allow charges for social security, medical insurance, retirement savings, and so forth, in addition to their salary limit, and some include them within it. Very few disallow them. If the application materials are not clear, ask the program officer how this information should be displayed on the budget page.

Capital Equipment

Agencies can be wary about these expenditures. You can't do research without the right equipment, of course, but be sure that you fully justify your choice of the make and model of each capital item. Do all you can to avoid giving the impression that you are writing this proposal only to buy pieces of equipment that your institution should have supplied as part of its educational expenses.

If your institution has comparable but inadequate equipment, demonstrate that you are familiar with the equipment on hand (best done by citing prior research and publications carried out with it), and then justify in detail your need for the newer, better instrument that you list. Then make sure to discuss discounts with the dealer, so that the line item doesn't cause reviewers or panelists to dismiss it as inflated because they paid 30 percent less for the same item.

A reasonable rule is to specify the simplest equipment that is compatible with the technical requirements of your research. Do not list the very top-grade equipment, figuring to bargain the agency down to a still luxurious second-best (a ploy known to field scientists as the "three-Land-Rover strategy": you ask for three, hoping to be cut to two, when you could get along with one—or bus fare).

It is rarely a good idea to low-ball the expense by getting an offer on used or refurbished equipment. Instead of winning points for frugality, it looks as if you are ready to risk the scientific aims of your research to save a few dollars. Every scientist can tell you stories of a piece of "free" used equipment that turned out to cost about as much as a new one, once the hardware and software upgrades, the installation and start-up, and the technician time to make it work are taken into account. List just the equipment you need, no more and no less. The process of deciding what you need will be a useful exercise in envisioning the research itself; it will often lead you to new ways of thinking about your science.

Major capital expenses almost always must be partly matched by your home institution, so be sure your provost, dean, and department chair are in agreement and that they (or your own start-up allowance or outside funds) will provide the required match. Most universities have funds specifically for this purpose, because it is cheaper than buying new equipment for their faculty or upgrading the old stuff.

Administrators are often receptive to such requests, especially if you can demonstrate that a number of other faculty are also interested in using the equipment.

Approach your meeting with the administrator as you would a major sales pitch. Bring two or three competitive estimates for what you want to buy, so it is clear what you want to purchase and how much it will cost. Bring a brief, one-page memo to leave with the dean, describing the equipment and what it can do, as well as where the equipment will be housed, how it will be maintained, and a list of the other faculty members or research programs it will help to support. You might also attach brief letters of support from your colleagues who are on your list of interested faculty.

This is also an excellent place to leverage small grants from local industry. For half or a third of the cost of a piece of important gear, a corporate sponsor can put its name and logo on a piece of equipment as if the sponsor had bought the whole thing.

If the proposal form and guidelines do not spell out how much of the cost of capital equipment should be shared by the institution, call the program officer and ask. A few agencies maintain a policy of deliberate vagueness on this question, so be sure you know before adding it into your proposal.

Travel

There are two kinds of scientific travel: conference travel and travel that is an integral part of your research, like fieldwork or trips to other locations where you can have access to specialized equipment. Most agencies allow you to budget some project-related travel to scientific meetings. That is one way in which their goal of having your research become widely known is met. However, some put a severe limit on this category of travel. If your fieldwork or meeting travel takes you outside the United States, check with the agency to see if you are restricted to traveling on U.S. carriers (as is the case for NSF, except under special circumstances).

If your research involves a cruise, work in the field, or use of a large facility like a linear accelerator that is not on your home campus, you will have expenses that can be charged to the grant to cover the cost of getting to where you need to go *and* staying there for some period of time. Estimate transportation costs at your institution's mileage rate, by getting fare information, or by calling various transportation carriers. Ask someone at the off-site location about inexpensive accommodations.

All agencies understand that fieldwork can be very expensive, particularly if it involves getting to remote areas and supporting yourself and students there. But be realistic. Reviewers and panelists for field-oriented research are field-oriented scientists themselves, and they know what things cost. Don't set out a carelessly estimated number.

Indirect Costs

The keenly watched category of indirect costs reimburses your institution for the expenses it incurs by maintaining the facilities that support your research. A few years ago, some institutions were found to be careless in the way they spent and accounted for indirect costs (which included the costs of maintaining a yacht), and they suffered the financial and public relations consequences. Since then, most agencies have considerably tightened their guidelines for budgeting indirect costs. Read the agency guidelines carefully, and note that some agencies disallow indirect costs entirely.

Because this is money that you will never see (it goes directly to the institution), it adds to the size of your budget without directly benefiting your research. As such, it can seem like a bad deal, forcing you to ask for more than you want for your own work. However, your research work does cost your institution money (for example, for heat and support staff). Indirect cost charges that your grants bring in are welcome supplements to administrative begging to meet these expenses.

Thus, good faculty citizenship, particularly at large research institutions, obligates you to do your share toward bringing in indirect costs. As a graduate student, part of your scientific education was underwritten by indirect-cost funds that came in from all the faculty. Now it's your turn to help the next generation. And even at institutions less focused on research, such as PUIs, indirect costs exist, and money to pay them is always welcome.

Expendable Supplies and Miscellaneous Services

This category is the one that most resists realistic planning. It certainly includes research supplies and expenses, such as any charges for computer time, reagents, laboratory animals and their care, expendable glassware, and external services such as per sample analytical expenses, slide preparation, and age dating. A host of small, hard-to-predict expenses like ink or toner for your printer, postage, small items of hardware, pump oil, repairs to equipment, swabs, gloves, and questionnaire preparation, printing, and distribution also settle into this category.

Although it is hard to predict these expenses, you must make a good-faith effort. After you've written out a spreadsheet that includes all the ancillary costs you can think of, take it to colleagues and your sponsored programs office and ask for experienced advice. In the end, the proposal will list a few large sums, not your item-by-item spreadsheet. So when you have done your best, don't be shy about rounding up to cover the things you undoubtedly overlooked. No one will actually believe a "Miscellaneous" budget line that adds up to $19,104.76 instead of $20,000 in any case, so you might as well give yourself a small margin for inflation.

Cost Sharing

Some agencies, particularly those in the government such as NASA or the Department of Energy or any number of state agencies that also do a lot of business with contractors, may require that your proposal provide some amount of cost sharing. This is generally a program-specific requirement and can be anywhere from 10 or 20 percent of the grant total, up to a one-to-one match. This allows the agency to show its constituents (generally politicians) that it is leveraging taxpayer money.

What a cost share means to you is that if you want to carry out a research program that costs $100,000 a year for three years, you (or your institution) will have to come up with from $10,000 to $50,000 a year. You may well ask, "Where is this money going to come from?"

Programs that require a cost share generally specify what can be considered as such. Only rarely is cash required. Often it is money that is already being spent. For example, if you are applying for a federal grant and you are a state employee (such as a faculty member at a state university or college), you may be able to count a fraction of your salary and fringes, possibly including institutional overhead charges, as part of your cost share. If you can get a corporation to donate a piece of equipment or defray its purchase price, this can also potentially be listed as part of the cost share. If your department waives the tuition of the graduate student working on the project or if the student is on scholarship from a source other than that from which you are requesting the money, this may be considered part of the cost share. Be alert to the possibility of in-kind matching by the institution that might be creditable without actually costing your institution cash. Sometimes institutions waive some or all of their indirect costs to make up part of the cost share. Because indirect costs have become a very important source of income to research universities, yours may be reluctant to forgo them. But it is worth bringing up as a possibility. Exhibit 4.2 gives an example of a suggested cost share on a proposal.

Coming up with a cost share is in many ways more an art than a science, and when constructing one, you need to go with a list of possible cost-sharing ideas to your contracts and grants office and be guided by the experience of the staff there. They generally know what can and cannot be considered for any particular agency. They can be invaluable in helping to construct something that will work for you and allow more money (in the form of overhead) to flow back into the university.

Subcontracts

One additional budget line you might encounter is a subcontract. This occurs when work related to the research will take place at some other laboratory or at some other location—for example, to contract for a ship for an oceanographic cruise or to pay colleagues at another institution to run a large suite of samples for

Table 1. Proposed cost sharing for equipment items, in dollars.

Item	NSF Contribution	Non-Federal Match	Total
1. iBook computers	$22,916	$22,916	$45,832
2. G4 PowerMac	3,332	3,332	6,664
3. Digital Camera	8,498	8,498	16,995
4. Projector	2,000	2,000	4,000
5. Imacon Scanner	7,300	7,300	14,599
6. IPTK Software	1,563	1,563	3,125
Total	$45,607	$45,607	$91,215

Exhibit 4.2. Proposed Cost Sharing for an NSF-CCLI Proposal
Source: University of Richmond, "Laboratory Investigations Using Quantitative Microscopy" (G. Radice, Project Director), 2001. Used with permission.

you in their laboratory. Not all agencies or programs allow subcontracts, so ask the program officer or check the program literature before you include one as a budget line item.

Generally, but not always, money for subcontracts is handled by the institution where the PI of the grant resides. Because the work is not being done at the PI's home institution, only a portion of the subcontract is generally subject to indirect costs. For example, at some universities, overhead is charged on only a fixed portion of a subcontract. The rest of the money in the subcontract is considered "overhead free." This convention varies from institution to institution, so if you are including a subcontract, check with your sponsored programs office first to see how your institution handles it. Occasionally, the agency will pay the subcontract directly to the subcontractor, but that is also something subject to different agencies and programs. When you can't determine from the agency's literature how a budget item should be portrayed, call the program officer.

The Budget Justification

Although budget justifications are not required by all agencies, it never hurts to attach a written description of your budget explaining the charges in plain English. This is another place not to let questions fester in the minds of your reviewers. Nothing seems to attract the interest of reviewers like an unusually large sum of money in a category that is inadequately described or a budget with wildly different levels of funding between years. I am not talking about a nickel-by-dime account of every cent you spend, but in your budget justification you can briefly tell the reviewer about your budget and what it entails. It is here that you explain that the first-year budget is high compared to the second because you are taking an early start date and will be funding a graduate student for two summers on the money. It can also be the place where you show that your second-year supplies and expenses budget is high because you will be buying platinum for catalytic substrates.

The budget justification is also where you explain what any institutional cost sharing consists of, and where you are able to leverage any other funds or services that will allow additional work to be performed on the project without incurring additional expenses to the grant. Exhibit 4.3 sets out the justification for the budget presented in Exhibit 4.1.

Remarks on Budget

Item 1. (P.I. Summer salaries). The amount requested is less than 2/9 of the P.I.'s academic-year salary, and is interpreted to account for two months of full-time work per year at $2,000/month. In reality, more than four months of full-time work will be spent on the project during the duration of the grant period.

Item 2. (Student summer salaries). The Principal Investigator manages a discretionary fund, up to $1,000 per year of which may be spent upon his own research. This sum will be used to supplement the amount requested from PRF to provide two 10-week student stipends at $200/week. In addition, St. Andrews Presbyterian College will provide housing for these students free of charge during their work on this project.

Item 3. (Supplies and services). Will largely be spent for thin section preparation and x-ray diffraction authentication of mineral samples, and for minor costs of assembling reaction vessel components. Such items as common chemicals, pH meters, bath thermostats, etc. are available to the PI at St. Andrews, as is use of SEM, of image recording and processing equipment, computer, and software for stereogrammetry. About $200 per year is estimated for SEM maintenance items (filaments, gold wire, etc.).

Item 4. (Capital equipment). The largest single item is a petrographic microscope (estimated cost $6,000). Although a few petrographic microscopes are available at St. Andrews, these are obsolete models donated or on loan from nearby universities, and are no longer capable of providing the high-quality, strain-free images and interference figures that will be required in the early stages of locating and isolating minerals from field samples. For a relatively minor cost above that of a "baseline" model, a trinocular instrument that can be used for teaching (projecting video images) can be obtained. The remainder of this item is planned for a peristaltic pump, and a Teflon reaction vessel.

Exhibit 4.3. Justification for Budget in Exhibit 4.1
Source: St. Andrews Presbyterian College, "SEM Stereogrammetric Studies of Etch Pit Growth Rates in Weathering Reactions" (T. Blackburn, Principal Investigator), 1988.

Time Lines

Some agencies, particularly those with a history of engineering like NASA, the Department of Energy, or Department of Defense, may require that you specify a schedule that shows each of the major tasks in the research program, when each will take place in relation to the other tasks, and what the expected milestones will be. This schedule is like the budget in that it can be only your best conscientious estimate. It is commonly presented in a graphical or tabular format, with time on the x-axis and tasks on the y-axis. Such figures can be easily generated using standard engineering program management software packages. They can also be created, with a bit more effort, by using any one of a number of good spreadsheet programs that will let you customize the format of the final document.

If the agency requires a timetable, do not forget to treat it just as you would any other figure, providing a brief discussion of it in the text and a good caption that describes all the symbols or colors used.

Supporting Information

You have told the agency what you want to do, why it's important, why you're the one to do it, how much money it will cost, and what your time frame is. And you have managed to squeeze it all within the imposed word or page limit. But you are still not finished.

Your Curriculum Vitae

Every agency has its own list of other things it wants to know about you, like your publication record, information about your institution or available facilities, or information about others who will participate in your project. To accomplish this, most agencies require some form of curriculum vitae (CV) from the PI and all other personnel above the status of graduate students.

The agency will also look for letters of commitment from identified collaborators or from those providing specialized samples, analyses, or data sets. Be sure you investigate carefully what needs to be included in your submission package. In most cases, agencies are very clear about these additional elements, and you will not need much external guidance. If the application instructions are not clear, however, do not hesitate to call the program officer and ask. Do not load your proposal with unrequested information. A good rule is to give the agency all the information it asks for and little else. (I have more to say about this in Chapter Six.)

Some agencies provide only skimpy blank spaces in their application forms for such professional qualifications as a list of your degrees, mentors, and publications. It is generally better, more readable, and perfectly acceptable to write "See attached CV" in those little spaces. This CV should contain the following information on you:

- Name and contact information
- Education and prior appointments
- Degrees received
- Majors and advisers
- Postdoctoral position and mentors
- Chronological list of your publications

The publication list should be clearly divided between peer-reviewed work (journal articles, book chapters) and nonreviewed work (such as abstracts or poster sessions at regional meetings). Even nonreviewed work is valuable if it establishes your credentials in a relevant technique or, for proposals to programs fostering undergraduate research, demonstrates a track record of working with undergraduate students.

The Institutional Setting

Most agencies require you to describe the institutional setting in which you are working—for example, your teaching duties during the proposed grant period and what support the institution offers for your research in the way of start-up funds, student support, departmental responsibilities, and so forth. Also, your institution may have characteristics (single-sex, underserved ethnicities, community outreach, special accommodations for disabled students) that coincide with broader social and program goals like those quoted from the NSF Grant Proposal Guide in Chapter Three. Your sponsored programs office may have the relevant institutional characteristics in efficient prose that you can simply insert into your proposal.

Ethical Considerations

Academic ethics can be a sensitive topic, yet it needs frank and open discussions of issues regarding what the norm is. Otherwise, newcomers to the academic ranks are left to grope their way through it (especially faculty who come from other countries and cultures) and hope they are not making any major mistakes.

Both newcomers and seasoned veterans in academe should share a common sense of what is acceptable and what is not, so we all play by the same rules. This holds true not only in protecting intellectual property, determining authorship, and use of borrowed samples or other materials, but also in certain aspects of proposal writing. The ethical issues include the names of persons you supply as recommended reviewers of your proposal and information on your proposals pending at other agencies.

Recommending Reviewers

Most agencies will ask you to recommend from three to a dozen scientists who are experts in your field and are competent to provide unbiased technical reviews of your proposal. This is an opportunity for you to ask the top scientists in your field to read your proposal and render judgment on your ideas. Do not miss such opportunities. Although you won't know the identity of the reviewers who are eventually selected, you are wise to take this chance to suggest people whose opinion you respect. However, you should also recognize that the top scientists in your field are often nominated by many other people as reviewers and are busy with other projects in any case, so they may not be available to provide a review for your proposal. Include younger scientists on your list whom you know to be well versed in your field and whose ideas you respect.

Avoid conflicts of interest. A person has a conflict of interest if he or she is unable, consciously or subconsciously, to provide an unbiased assessment of your work (good or bad). Of course, applicants are unlikely to request proposal reviews whom they know to be biased negatively toward the research. Similarly, no one should recommend a reviewer who might not be able to untangle personal opinions about the author from the science that needs to be evaluated.

Your graduate school adviser or a recent postdoc mentor may be the person best qualified by expertise to render judgment on your research ideas, but that person would have a conflict of interest in doing so: your success in getting grants would reflect well on your mentor. The same is true of *any* member of the faculty of institutions at which you were enrolled or where you served as a faculty member, of the other members of your past research groups, your own past students, and most of the people who have coauthored publications with you. These are all people who have already formed an opinion about you as a person and will have a difficult time separating what they know about you or your personal situation from the science you are proposing.

Part of the job of the program officer is to sniff out these conflicts and to eliminate the conflicted person from the list of possible reviewers. Program officers tend to get very good at this, and when they succeed, it doesn't make them think well of the researcher submitting the proposal. When they don't detect the connection and they do send a review request to someone with a conflict of interest, then *that* person will not think well of the proposal submitter. When the situation is brought to the light, it may delay a decision on the proposal and even ensure a negative decision. You want as unbiased an opinion of your ideas as you can get, even though sometimes it doesn't feel that way when a review comes in. Grit your teeth, and play it straight.

You can also request that certain individuals not be asked to review your proposal, and these requests will be respected. Nevertheless, you should not do this unless you have a very good reason, which you have discussed with an objective third party whose judgment you trust. You might consider this course of action when your proposal espouses a controversial point of view, to which you know for a fact that a possible reviewer is hostile and unlikely to offer an objective review. It is also warranted when you believe, again on good evidence, that a possible reviewer cannot be trusted with confidential information or ideas presented in your proposal. This request should be put in a communication—either a separate e-mail or letter or both. It should not be put anywhere in the proposal, because all the reviewers and the members of the panel will then see it.

Pending Proposals to Other Agencies

On any grant proposal, most agencies will ask you to list your current grants and all other proposals you have under consideration. When answering this, be sure to make accurate statements about the extent to which proposals to other agencies overlap with the one you are submitting and to keep this information updated with each agency. You gain much more by being honest and transparent about duplicate or overlapping proposals than any temporary benefit that you might be tempted to realize from trying to get duplicate funding. Agency grants are public information, and program officers talk to other program officers and have long memories. (I have more to say about multiple proposals in Chapter Six.)

Strategic Considerations

Most agencies have guidelines that govern the proposal and review process, such as length limitations and deadlines. Though these have nothing to do with science or money, they can be crucial. Here is a survey of some of the most important considerations.

Add-ons and Appendixes

There will inevitably be things that you want to tell the agency or the panel but the application doesn't provide space for, and you will be sorely tempted to add these to your proposal (for example, reprints or preprints of articles you've written in the field, supplementary data tables, or accolades from your students or your dean). Here is a simple rule that will help you decide whether to attach this kind of extraneous material: Don't.

If your recent article in the field was voted the Best Paper at the national meeting of the Planetesimal Society, you will, of course, cite it in your narrative and list it in your bibliography, with exactly the same level of fanfare as you give every other paper that you cite. Reviewers familiar with your field will also be familiar with your excellent paper. If you attach a copy of it as an appendix so reviewers can admire it up close, they will not thank you for giving them more to read. Your reviewers probably will ignore it if the program officer hasn't already discarded it. Program officers, reviewers, and panelists see added documents as just one thing: an attempt to get around the narrative length limit. Do not attach appendixes to your narrative unless they are specifically requested by the agency.

Length Limits

All agencies set a limit on the length of proposals, expressed in terms of the number of pages or as a word count. If they didn't they would be buried in proposals as thick as telephone books. Limits on length can be fifteen or twenty pages (four thousand to six thousand words) or longer, or as short as two to three pages. Some agencies include figures and tables in the limit; others do not. Be sure that you understand the length limitations of the agency to which you are applying, and then respect them.

Also respect all specified margin and font requirements. Your goal is to make your proposal easy for reviewers and panelists to read. Do not try to overcome a page limit by widening your text so it covers the page from edge to edge. Remember that your best friend is a proposal that is easy on the eyes. It dramatically improves the attitude of the person reading the proposal, as well as the probability that everyone will read the entire document.

The page limitation requirement is not something you should be calling your program officer about unless you find the stated limit incomprehensible. Any proposal can be shortened, and indeed most would be greatly improved if they were. If your narrative section is over the limit and you can't seem to get it under, let it cool off for a day or a week before going on a search-and-destroy mission for excess sections, paragraphs, clauses, prepositional phrases, and words. You will be amazed at how much you can eliminate and condense (usually from the introduction or the background section) after you take a break from your proposal for a few days. (The question of writing clearly and efficiently is addressed in Chapter Five.)

Length limits can be frustrating, and you might swear that there is not another comma you could cut from your over-length narrative, but these limits are your friend. They force you to think succinctly about what is really important in what you are proposing to do and to write clearly and tightly about it. It is said in the business world that if you can't put your entire presentation on one slide in three bullets, then you might as well forget about getting your boss to okay your project. Scientific proposal writers should take note. Clear, efficient writing is much easier for reviewers and panelists to understand, and it makes your proposal much more competitive.

Deadlines

Here is another source of stress for all proposal writers. Some agencies maintain sharp deadlines, after which your proposal is rejected without review (as is the case of the NSF, whose deadlines are to the minute). Others receive proposals year round, deciding on short notice when to stop accepting proposals for the current review cycle, and assigning all subsequent proposals to the next. Under this system, no proposal is ever rejected for missing a deadline, but consideration of it may be delayed for three to six months.

Your strategy for meeting deadlines or cutoff dates will be to beat them by not less than one week, but preferably by several weeks or more. Here are six reasons:

1. You will invariably find at least one mistake, typo, or blunder after you have finished the proposal and it has rested on your desk for a few days. Astounding numbers of proposals are submitted with simple arithmetic errors in the budget, with paragraphs that end in midsentence, with whole pages missing, or with no investigator CV or bibliography for the narrative. If you finish a proposal and submit it right before the deadline, there's nothing you can do to remedy any of this.

2. Remember that your proposal is actually being submitted by your institution, not by you. Therefore, it must be signed by at least one administrator who is authorized to commit the institution to the financial contract you are proposing. Those administrators have a vexing habit of going on vacations and administrative retreats just when you need them the most.

3. At many institutions, the sponsored programs office and possibly an in-house photocopying facility will also have to be involved in assembling and mailing (or electronic submission) of proposals to the agency. These are the main customer base for those familiar posters that say, "You want it *when???*" And "Failure to plan ahead on your part does not constitute an emergency on my part."

4. You must insist on checking the work of these offices personally. The fact that you submitted a perfect copy of your proposal to them for signing, copying, and mailing does not mean that it will leave their hands in perfect order or on time. Late submissions and imperfect copies with smeared or missing pages may be their fault, but they are your responsibility. You can't meet that responsibility if you don't give yourself enough time before the deadline.

5. Agencies that use electronic submissions and sharp deadlines (for example, NSF via its Web-based proposal management system, FastLane) often find that their servers cannot adequately handle the volume of Internet traffic that comes on the last day or two before a deadline. If you wait until the last minute with these agencies, you run a real risk of being frozen out of the system until after the deadline has passed. The agency may (or may not) be forgiving of this, but you don't

need the aggravation and heartburn of staring at that hourglass on your screen, waiting for something to happen as the minutes tick by, closer and closer to the deadline. As of this writing, there are about 250,000 registered users of FastLane (Bradley J. Stith, associate professor of biology, University of Colorado at Denver, remarks following his lecture at the Council on Undergraduate Research Proposal Writing Institute, Juniata College, Huntingdon, Penn., July 2002). Of course, they are not all submitting proposals in your group, but that's a big pool of competitors for CPU time on NSF's computers.

6. When the agency uses a rolling review cycle system, it may have guidelines as to the number of proposals it will send to any single reviewer in each cycle. If you submit late in the cycle, you are much less likely to get the reviewers you suggest, and your proposal may be deferred to the next panel meeting because the program officer could not get enough substantive peer reviews in the time remaining before the panel meeting.

The cure for all of these issues is simple: use a working deadline for completion of your proposal that is at least two weeks earlier than the agency deadline. Lie to yourself if you have to to maintain this fiction. When your proposal is submitted, perfect, and in plenty of time, reward yourself with a nice lunch. Then start working on the next proposal.

Summary

Writing a good budget can be as time-consuming as conceiving the scientific ideas it is meant to support, but it need not be as difficult. Build your budget from the bottom up, supporting your numbers by surveying typical costs for each activity or object and taking into account regional salary differentials. Use successful proposals (such as those on the Web site for this book) as models.

Unlike the scientific part of your proposal, the budget is very much the business of your institution's administration. Use the institutional experience of your sponsored programs office, and work with your department chair, dean, or provost to put together institutional matches and other cost-sharing information.

Make sure that you are familiar with agency guidelines (prominently noted on most agency budget forms) for each budget category and for the overall total.

Never sign a proposal budget that you don't understand or that you can't account for, line by line.

Be aware of ethical and strategic considerations: avoid suggesting reviewers with a conflict of interest in your work or your career; keep all agencies informed about parallel or overlapping proposals; and avoid attaching add-on documents like article reprints and preprints and letters of recommendation by third parties. Respect length limits not by omitting ideas but by writing about them efficiently. Manage deadlines by scheduling your writing to beat them by a wide margin.

Achieving "Excellence"

Sometimes the process of avoiding sin can force you toward virtue, or at least put you within striking distance. One approach to writing research proposals judged as "Excellent" might simply be to avoid writing bad, fair, good, or even very good ones. In this chapter, you will revisit and reinforce some of the tips brought up earlier in the book; however, this time you will be looking at them from the point of view of the characteristics that tend to lead reviewers to assign a particular proposal score from "Poor" up through "Very Good" or better. I will take the somewhat perverse approach of starting with the "Poor" end of the scale and working up from there. Still, excellence is a positive characteristic that means more than just faultlessness. I'll pinpoint key ingredients that can transform proposals judged "Very Good" into ones deemed "Excellent."

I am starting at the dank and gloomy bottom of the heap. I will first be describing problems that may seem unlikely or extreme to you; nevertheless, they are things that program officers see all the time. As we advance through the chapter, the mistakes will become increasingly subtle; some may even begin to look like good ideas.

Train Wrecks: Not Writing "Poor" Proposals

Readers with teaching responsibilities may have noticed a striking truth about student behavior: most students who receive the lowest grades in any course almost always do so, in effect, voluntarily. They "volunteer" for the bottom of the grading curve by failing to attend class, failing to turn in assignments, and so forth. Rarely does an earnest effort, even a wrong-headed one, result in an F. In just this way, my experience has been that proposals that rank at the very bottom of a cohort have pretty much "volunteered" for that fate. Let's look at some of the most common ways this happens.

Wrong Agency, Ineligible Proposal

There are some proposals that don't even get to a "Poor" rating because they are never sent out for review. These are proposals that were sent to the wrong agency or wrong grant program because the writer did not read the literature carefully and speak to a program officer before submitting it or did not address the goals and objectives of the agency. The proposal may also have been submitted late (and thus, for some agencies, rejected without review) or submitted on obsolete forms. The proposal narrative may far exceed the word limit, or its budget may be larger than the maximum grant that the program offers. The proposal may have been submitted to a particular program by an investigator who is ineligible for it. Examples might be an NSF Experimental Program to Stimulate Competitive Research (EPSCoR) proposal, available only to researchers in underserved states, from a California institution; or a Cottrell College Science Award, reserved for nondoctoral departments, from a department with a doctoral program; or a midcareer investigator applying for a program that gives only starter grants. Difficult as it may be to believe, these and similar violations of basic guidelines crop up during every round of competition by every agency.

Weak Science

Some (though a very few) proposals get ranked at the bottom of their cohort because the science proposed is plainly inadequate. Their premises fly in the face of the laws of thermodynamics or quantum physics or the conservation of momentum. The equations don't balance, or the surveys contain leading questions or gender insults, or the analysis of a text requires that Chaucer was a fan of Debussy and capable of Unix programming.

Most of the time, the errors are not so obvious. They are more likely to hide in the background, never mentioned explicitly in the narrative. Better examples might be a chemical synthesis that requires a mechanism that is inconsistent with a known rate equation, or a transfer of culture between ancient populations whose apparent relative ages depend on which dating technique one uses. These oversights can get by researchers when they are so familiar with the phenomena that they don't take the time to think through all of the theoretical underpinnings, step by step, the way a good peer reviewer will do.

That is why it is imperative that you find colleagues whom you trust to serve as preliminary readers for your proposal and ask them to check your logic. This, by the way, is one of the many reasons that you should always work to a very generous deadline. No one can provide a searching critique of your science if you say, "Would you be willing to help me out by reading this proposal and getting back to me after lunch?"

Out of Date

A third, and slightly more subtle, way of landing in the bottom 10 percent of the proposals in your cohort is to slack on the work of surveying the state of research in your field, and thus to submit a proposal for work that has already been or is currently being done. This is why I made such a point of the preliminary survey in Chapter One.

This happens more often than one likes to think. Scientific research is a communal activity, and that community is, in this way, more like a Samoan village than a faceless metropolis. If some piece of research has already been done and published before you get around to proposing it, someone in the reviewer community will know about it and will doom your proposal with a few words and a literature citation. (In fact, there's a good chance that your proposal will be sent to that prior worker for review. And even if the missing reference would have been supportive rather than duplicative of your idea, missing it makes you look careless.)

Cavalier Submissions from Top Scientists

There is an old saying, "If you're famous enough, even your bad proposals get funded," but it is rarely true. Once in a while, a program officer will receive something that looks like a proposal, but reads like a transcript of cocktail party conversation, written on the back of a napkin and signed by a famous scientist who evidently believes that careful writing is no longer necessary. It is never a pleasure to send such proposals out for review or to read the baffled and diplomatic opinions that come back.

The Pallid Busts: Not Writing "Fair" Proposals

A glance back at Exhibit 1.1 in Chapter One shows that in talking about the sins that produce "Fair" scores, I am still speaking of a rarity—proposals that fall into the lowest 10 to 20 percent of the cohort. These sins are paler than those that provoke "Poor" scores but are still relatively easy to spot and avoid.

Obsolete Research

This is not quite the same as research that has already been done. Rather, it refers to areas of inquiry that have been abandoned for lack of evident payoff. No one wants to believe that the research they love is something that very few others care about. But if the most recent entry in your narrative bibliography is ten or twenty years old, that's telling you something.

This is not to say that fresh approaches to old, unsolved problems are not competitive. If your research will finally solve the origin of Liesegang rings, develop a less toxic smallpox vaccine, or reveal the phylogeny of the Loch Ness monster, you have a great chance of attracting funding. But you had better be talking about these old problems from a contemporary perspective with a well-defined theoretical model, the best available tools, some preliminary results, and some testable hypotheses.

A subcategory of this group of endeavors involves hot, relevant, and contemporary problems approached with obsolete methods and tools. Some examples might be using a scanning electron microscope to investigate surface structure when atomic resolution is needed, doing chemical engineering reactor kinetics without detailed computational modeling, or investigating recent tectonic displacements without a global positioning system. If you have promising ideas for working in these areas but no access to the best instrumentation, you should consider collaborating with someone who does. That is far better than going it alone with a proposal that will be downgraded because you are attempting brain surgery with stone knives.

Lame Logic

You see lame logic in fishing expeditions (projects whose premise has yet to be proven) and in quantitative studies that are too error sensitive to be reliable. For example, any technique that requires measuring a small difference between two large quantities always greatly magnifies experimental error.

"Edisonian" projects also fall into the lame logic category. These are projects whose success depends on trying, in sequence, a number of equally likely techniques, as Thomas Edison did when he tried and discarded hundreds of candidate materials for filaments before his light bulb stayed on.

A third example of lame logic might be a study that uses the weathering rate of a mineral to deduce both exposure times and temperature history, two unknowns that can't be isolated from a single measurement. All of these, and any study whose logical structure boils down to, "If A is true, then B and C will follow," when A is speculative at best, is going to be rated "Fair" at best.

Projects with Limited Impact

It is possible for a research project to be well designed and theoretically sound, with testable hypotheses and robust logic, and still not amount to anything because the contribution it will make to science is too modest. In such cases, either a negligible incremental advance is proposed for an important problem (what I call the "Big Bing") or a complete solution is proposed for a trivial problem (the "Small Bang").

Examples of Big Bings might be to measure equilibrium states or rate constants for two or three chemical reactions without tying these results to some larger vision of reactivity. "Effects of . . ." projects are also in this category—for example, "Let's shine green light on dandelions and see if they still bloom as well as those getting blue light." There is nothing wrong with well-measured physicochemical constants or well-controlled horticultural experiments. These might make excellent three-week projects for an undergraduate laboratory course and in the process teach the students something about the nature of research. But they don't promise as much bang as comprehensive studies, like determining hydrolysis rate constants for a series of substituted benzoate esters, to name a famous example (Hammett, 1938), or making a detailed spectral analysis of photosynthesis on the molecular level.

Small Bangs are made when small problems are completely solved. Many projects in environmental chemistry fall under this category when there is no carry-over from one field area to others or no more comprehensive conclusions promised than, "Well, we looked in this watershed, and this is what we found" (Blackburn, 1973). These may be perfectly sound studies that leave no stone unturned, but the researcher generally makes no attempt to relate the results to larger issues and questions.

In both of these cases, there may be nothing at all wrong with the science. Nevertheless, the proposals will attract scores of "Fair" and lose out to proposals that promise more scientific impact for the dollar.

Unanswered Logistical, Safety, and Ethical Questions

Time and again, field-oriented proposals have failed to attract funding because a reviewer or someone on the panel is familiar with the field area and knows that it is subject to control by established researchers, indigenous populations, or even warring groups who are touchy about incursions. If that is true of your study, your proposal must specifically address the problems of access and personal safety. The same is true of proposals that seem to risk exposure of the investigator, students, or the work's immediate environment to dangers such as toxic substances, radiation, volcanic eruptions, or pathogens.

All safety, environmental, and logistical details must be addressed and solved well before you submit the proposal, and you need to describe their solution in the proposal text. Proposals that appear to ignore the ethics of experimentation with humans and animals, negative environmental impact, fieldwork access and costs, and experimenter safety are doomed to the lower ranks of the proposal pile.

The Enemy of the Best: Not Writing "Good" Proposals

We now face some cases of viable ideas done in by systemic, but fixable, problems. Because they occur in proposals whose basic scientific idea is sound and ought to be carried out, they are particularly crushing and hard to understand when your reviewers call attention to them. If you find yourself with a handful of "Good" reviews, be ready for some hard work, but also be encouraged that you have come up with a fundable idea that may need reformulating but should not be given up on.

The Researcher's Ability to Do the Work

Researchers at the beginning of their careers or at relatively unknown institutions face a hurdle that well-known researchers at top universities may once have faced but no longer do: the question of their ability to carry out the proposed work. If you are in either of the first two categories, your proposals will be less immediately convincing now than they will be later, when you have become a known quantity.

The merit of a proposal is not independent of the human and institutional setting in which the proposed work will take place. (For an official statement of this fact, turn back to the NIH review criteria in Exhibit 3.1. They address the investigator.) The same proposal that might be graded "Excellent" coming from an associate professor at a top university might get only a "Good" (as in "good try") from a new assistant professor at a small institution unless the proposal provides solid evidence, in the form of institutional support, prior publications, preliminary results, or co-PIs of high repute, that money spent on this research will buy published results. This may not seem fair, but it is a factor until you prove yourself. And I have just listed what you can do to overcome it.

Poor Organization and "Lack of Focus"

"Good" proposals include those that have no abstract or summary, even where the application instructions don't seem to require it. They include those whose abstract is not an exact map of the narrative, introducing ideas not found there or failing to mention significant ideas that are featured in the narrative. Proposals whose structure and organization are difficult to understand, even with the best will, are likely to earn a "Good." Even omitting such trivial details as numbering the pages so that reviewers can easily keep the document in order and refer to passages of interest can hurt you. If you introduce ideas more or less as they came to you while writing, without reference to logical organization, you are also likely to wind up with "Good" reviews or worse.

One of the most common complaints about proposals from beginning researchers is that they "lack focus." This is a reviewer term for a proposal that promises too much (see "Promising Too Much" in Chapter One). Reviewers are generally familiar with how much work can be done in two or three years with a few undergraduates or with a graduate student and half of a postdoc's time. They also know about how many cleanly testable hypotheses (four at the outside; most say two to three) make up a cohesive and satisfying unit of research funding. If your proposal reads more like a laundry list of intriguing ideas than like the blueprint for a well-run, successful, limited campaign, you are likely to be given a "Good" pat on the back and told to go back and focus your thinking.

Too Timid

This is the counterpart of the proposal that promises too much: it promises too little, and not necessarily in the way of scientific results but in the way of too little intellectual development beyond the work you did as a graduate student or postdoctoral fellow. Reviews of this sort are likely to characterize your proposal as "derivative" or "a minor extension of Dr. X's graduate [or postdoctoral] work in Y's lab."

You get here by writing a proposal that is little more than Chapter Ten of your nine-chapter thesis. Reviewers care about that, because they are conscious of the personal and financial resources that have been invested in your scientific education, and they believe that the scientific community is entitled to more than just a modest return on that investment. Because a proposal with this problem is likely to show good command of the research field and a high likelihood of being competently carried out, it is likely to draw a score of "Good." Reviewers who see little new thought going into it are unlikely to score it higher.

The Opaque Budget

In some budgets submitted with otherwise very sound proposals, the origin of and need for each budget line is not clear. These include such items as "Field Work: $15,000," when the narrative seems to describe a two-week effort not far from your home territory, or "Assistants: $8,500," when no inkling is given as to who these people are and what their duties or qualifications are. (If you mean graduate students with the title of research assistant, list them under "student support.") Even such a reasonable and transparent request as "Atomic Absorption Spectrophotometer: $25,000" is a mystery if no information is offered as to its capabilities, make and model, or the extent of institutional cost sharing.

Reviewers sold on the project and ready to award and justify an "Excellent" rating will downgrade your proposal to "Very Good" or "Good" when faced with a mystery budget. It gives the impression that you haven't thought very carefully about what you really need to do the work and how much it will cost. (For a complete discussion of budgets and budget justifications, see Chapter Four.)

Inflated Prose: Don't Utilize "Utilize" When You Can Use "Use"

Turgidity is the enemy of clarity, and thus your enemy as well. Beginning researchers (and even some who have been in the game a long time) who want to appear serious in their thinking and writing sometimes reach for a "high" and serious style of writing that multiplies syllables, stretches sentences, and sometimes defies interpretation. Passive constructions are "utilized" (as in the first half of this sentence) where you should use an active voice. If you follow this path, you begin to sound like Sherlock Holmes's pal Doctor Watson: stuffy, formal, and a little dense. Exhibit 5.1 provides some examples, followed by their plain-speech alternatives. The problem here is not so much formal language as inflated, marginally understandable prose.

Reading proposals is hardly anyone's idea of fun, but readers know how much is at stake and do not take their job lightly. The whole reviewing enterprise, and thus our best attempt at the fair distribution of scarce research dollars, depends on scientists who are willing to do this work carefully and without pay. Don't abuse that social contract by making your proposal a chore for them to read.

In addition, by obscuring the scientific content of your proposal, inflated prose makes assigning an accurate score harder for the reviewer. Your proposal may get

---

Inflated: "The experimental protocol is broken into two sequential parts, A and B."

Better: "The experimental protocol consists of A followed by B."

Inflated: "Product produced in Scheme 1 will be utilized in Reaction 4a."

Better: "Reaction 4a will use the product of Scheme 1."

Inflated: "Background spectra will be obtained in the solvent alone, in the absence of the analyte."

Better: "Background spectra will be obtained in pure solvent" or "Analyte spectra will be compared to solvent spectra."

Inflated: "The most utilized arachnicides employed in animal colonies have been silica-based C_{20} polycyclics, comprising about 70% of the arachnicides."

Better: "Silica-based C_{20} polycyclics comprise about 70% of arachnicides used for animal colonies" or "About 70% of animal colony arachnicides are silica-based C_{20} polycyclics."

Exhibit 5.1. Examples of Inflated Prose and Better Alternatives

an unfairly low score, or it may get a high score (since no one wants to be unfair) that is not justified by an informed and cogent narrative review, just because the reviewer was unable to sort out key ideas in your proposal. In either case, your ranking by the panel is likely to suffer.

A good rule for writers and readers alike is this: *unclear language comes from unclear thought*. If you have trouble expressing your ideas clearly, maybe you don't have them as clearly in your mind as you think you do. Here are a few brief guidelines:

- Write short sentences.

- Favor verbs over nouns, and both over flimsy adjectives.

- Replace passive constructions with active ones.

- Use short, vivid words like *use*. Avoid longer, inflated synonyms such as *utilize* that signify the identical concept. (See the difference between these two sentences?)

There are other ways for you to strengthen the presentation of your proposal:

- Divide your narrative into sections with bold subheads that announce the subject of each section. Good headings keep your readers on track and allow them to find where they left off reading, because it will probably take them more than one session to review conscientiously. Headings also allow readers to find the section of your proposal that they want to examine in more detail without having to grind through a forest of undifferentiated text. If you have two or more proposal pages in a row without a heading or subheading, you need to consider including one to let readers know what you are talking about now and where they are in the big scheme of things.

- Begin a paragraph at every logical spot, and indent the paragraph because that gives the eye something to check off while reading. Concise language will lead to compact paragraphs and frequent reader relief.

- Avoid too many acronyms, unless they are all commonly used and accepted (TEM for transmission electron microscopy; GC for gas chromatograph). This applies as well to the abbreviations you make up because you think it is too grueling (or murderous on your word count) to use the full term frequently. An example might be something like "mass dependent multicomponent dendrimer copolymers," which you might then refer to throughout the text as MDMDCPs. Readers with poor short-term memory will grit their teeth over this. And if you define more than one of these, readers will be looking back through the text for those definitions as well. It is much better to use the full term when it is needed, even though you get tired of typing it. Your job when writing your proposal is to make as little work for the reviewer as possible.

- The same is true about technical jargon. This is not to say, of course, that you should look for ordinary-language substitutes for well-understood technical terms. When it makes a difference, call things by their right names. Just don't show your erudition by calling a spade a manual detritus displacement implement (or MDDI).

Using efficient, lean prose means that you will fit more comprehensible intellectual content into fewer words. You will have more room to do a thorough job of explaining your research ideas without bumping up against the narrative length limit. Most difficulties with word or page limits come from writing that inflates a description into a dissertation.

The cure for pompous language is not to write as if you were sending a postcard to a friend. Keep your language focused on the matters at hand, resisting the temptation to begin sentences with, "The other day, walking to class, I began to think about the Schrödinger equation, and . . . " or "As anyone can plainly see . . ." Don't refer to molecules, mathematical functions, or other inanimate things as "animals" or "beasts" or microorganisms as "bugs." This kind of breezy writing is a cousin of inflated writing: it puts useless words in the way of understanding. It also puts you in the center of the picture instead of the ideas you are writing about. *Always remember your audience*. Your readers are active scientists with long lists of things they have to do after they finish reading your proposal. They are not looking for entertainment or personal glimpses of you, your life, or your students. They are looking for quick, clean insights into exactly what it is you propose to do, why it is important to do that, and how you are going to do it. Provide those, and your excellent idea will escape the "Good" rating that follows any hint that you are not 100 percent serious and professional about your work.

Almost There: "Very Good" Proposals

The problems I discussed under the "Poor" category are fatal in even the smallest doses. You can no more be somewhat ineligible for a grant program than you can be a little dead. But the "Fair" and "Good" categories of mistakes do come in various sizes. Your proposal can be a *bit* unfocused, your budget can be *partly* mysterious, the scope of your aims can be a *little* too modest. An otherwise fine proposal that suffers from a very mild case of one or more of the "Fair" or "Good" flaws is likely to be rated "Very Good."

"Very Good" is a score that reviewers use to convey the encouraging message that the proposal is basically sound and potentially fundable but could be much better with a bit of rewriting or a shift of emphasis. Reviewers also use "Very Good" when they see nothing really wrong with the science but think, "It just didn't thrill me." This translates as, "It's mainstream science with only a minor promise of impact on future work."

To advance beyond "Very Good" to the really competitive ranks, you are going to have to let your proposal cool off for a few days before pressing on, another reason I recommend that you work with generous deadlines. When you come back to it, you need to read it very critically, looking for and eliminating all of the pathogenic little flaws you visited above.

And as you read through your proposal, constantly ask yourself questions and be sure you have answered them. Exhibit 5.2 provides an example of this.

Write out explicitly the logic of your hypotheses and the experiments that will test them and the impact of your work on science, making sure that you have all the ethical, logistic, and environmental bases covered. This will not be a simple job, like checking things off a grocery list. It will take you days. If you do it right and fix all the flaws you find, you will have a better than "Very Good" proposal. But you must leave yourself time to do it.

From the Heart: Writing "Excellent" Proposals

I do not mean to imply that a proposal will be rated "Excellent" simply because it is free of significant flaws, though that is a necessary condition. But assuming that you have an important, valid scientific idea and that you can honestly say to yourself that your proposal shows none of the problems identified so far in this

Q: Am I really eligible for this program?

A: Yes, because I called the program officer and confirmed it.

Q: Have I drilled down into the fundamental basis of my experiments and observations, and do I know for sure that there are no violations of the accepted canons of my field hidden down there?

A: Neither two colleagues I trust nor I found the least problem with my science.

Q: Is my budget clean and transparent?

A: Yes, and the Sponsored Programs Office thinks so too.

Q: Have I discussed the impact of my work on the field?

A: Yes, on pages 11–12 of the narrative.

Q: Have I shortened every sentence and stamped out wordy, passive, jargon-laden prose?

A: Yep. And replace "wordy, passive, jargon-laden" in that question with "inflated."

Q: Is the work I propose clearly related to my previously demonstrated interests and competencies, and does it show evidence of a significant advance beyond them?

A: Yes, yes, yes!

Exhibit 5.2. A Conscientious Applicant's Self-Test

chapter, then you are on the cusp of excellence. Beyond this point, there are two features that will push your proposal into the promised land. The first (preliminary results) is relatively easy to explain and understand; the second (conviction) is less so.

Preliminary Results

Preliminary results are real laboratory data that you have produced in real experiments of just the kind you are proposing. This does not mean a one-afternoon study that produced a little pink foam in a test tube. But neither does it mean finished work that is already on its way to publication. By "preliminary results," I mean work that is probably well enough conceived and executed to justify a poster presentation at a regional meeting where submissions may be lightly peer reviewed, if at all, but where you have the opportunity to present and defend your work.

Maybe this work amounts to carrying out the first one or two of a planned dozen sequential steps; or a preliminary survey of the field area you plan to examine in detail; or a collection of clean, reproducible data that are critical to the larger project described in your proposal. Similarly, if you are proposing to use a substance in your research that is not widely available, you might demonstrate that you can actually make it in the purity that you are going to need to do the research.

If the grant program to which you are applying has as one of its priorities the encouragement of undergraduate research, it is doubly golden if that regional poster presentation has an undergraduate coauthor as the designated presenter because the undergraduates working in your lab were the ones who helped generate the data.

Preliminary results should appear in your research plan (as opposed to the Introduction and Prior Work sections of the narrative), in a separate and clearly labeled section so no one will miss them. This allows you to establish, in one blow, the existence and general nature of the phenomena you want to study, the basic soundness of the techniques with which you propose to study them, and your competence with those techniques. Often, the presence or absence of preliminary results makes the difference between an "Excellent" and a "Very Good" score.

You may suspect a double-bind here. If you need a grant to do the work, how can you do even part of the work without the grant? The answer is that you may have to work without extra pay for a few weeks during the summer. Sometimes you can compensate student assistants with academic credit instead of money. You may have access to enough internal money to buy a limited study. You may have to buy or borrow time on an instrument like the one you plan to buy with your grant. I'll talk a little more about this in Chapter Six, but the point is that you may have to be ingenious, and you may have to scramble, but the payoff can be substantial.

Conviction

The difference between writing "Excellent" research proposals and writing "Very Good" ones is not just in having an excellent idea and presenting preliminary results that demonstrate that you can carry it out. It certainly does not lie simply in not writing poor (fair, and so forth) proposals. It lies in your motivation for writing the proposal in the first place and in the conviction with which you wrote it.

By *conviction*, I mean your wholehearted belief and ability to convince the reader that the research you propose *must* be carried out, without delay, and that even if you were not the person to carry it out, you would welcome the news that someone else had done so and would await with eagerness the publication of the results.

For some scientists, this wholeheartedness exists almost from birth, and certainly from the time they enter the field. These people—given other conditions, such as an able (not necessarily genius class) mind and the bare opportunity to do the work—are, or will be, the leaders in their field and the winners of medals and prizes. But admirable as these people are, they are not the only winners of grants.

Others, like me, may be competent researchers, but they may also be committed strongly and primarily to other parts of their jobs: the careful and creative planning of lectures and laboratories, the give and take of teaching students, the hard work of faculty governance, the intellectual camaraderie of professional life. This group overall may include fewer prize winners, because for them, the motivation to write research proposals is tempered by other involvements and issues.

Some people write proposals because they recognize that their students deserve a chance for hands-on research as part of their education. For some, the motivation to seek grants may arise from nothing more inspiring than a feeling of being left behind by younger, more energetic colleagues, or even a note from a dean or department chair that prods them to do it.

If you are in these latter groups, the conviction that you are on to a must-be-done research idea may not emerge until you have gone through the rigorous discipline of surveying your field, probing the theoretical underpinnings of your idea, and pushing yourself to expand your intellectual horizons. It may not appear until you get one of those late-in-the-process ideas that you never would have gotten if you hadn't started writing with some idea, however partial and preliminary. Conviction may hit you as late as when you are writing out the statement in your narrative about the potential impact of the work you are proposing. In the end, though, it doesn't matter. In an excellent proposal, it will be there. Conviction cannot be checked off on a list, and it can't be faked. You will recognize it when you have it, and—here is the point of this pep talk—so will your readers.

Wholehearted conviction about the research you propose is an indispensable ingredient of powerful, grant-winning proposals. It will reveal itself in the thoroughness of your planning, the imagination with which you consider alternative models to your hypotheses, the specificity and care with which you budget your funding, the wisdom with which you choose the three or four really crucial ideas out of the basketful that you create, and the shrewdness with which you prioritize the others. Conviction will be recognized by the reviewers and panelists who will determine your proposal's fate because it will reveal itself in almost every sentence you write. And so will its absence.

Summary

We have looked at some ways in which promising ideas for research can lose their way and become poor, fair, good, or (only) very good proposals. At the low end of this sequence are fatal problems (scientific errors, obsolescence, ineligibility) that lead to "Poor" or "Fair" scores, and are virtually impossible to fix without starting over again.

"Good" proposals, winding up in the middle of the pile (but well short of the funding cutoff), are those that also need serious rethinking and rewriting. They are the proposals that do not establish clearly the importance of the project or the proposer's ability to carry it out and may have some flaws in methodology or approach. For many of them, this lack of clarity, completeness, and focus is reflected in inflated, wordy, opaque language. Sometimes, this opacity extends to the proposal's budget.

"Very Good" is often a reviewer's way of saying, "Good science, close to fundable, but not quite over the hump." These are proposals that are basically sound; they will probably not need a complete rethinking and rewriting, but they are weakened by mild cases of the same flaws that, when too serious, would have resulted in scores of "Good" or "Fair."

"Excellent" proposals not only avoid all the problems discussed in this chapter but also promise success by presenting preliminary results and by the clarity and conviction with which the writer presents the problem and describes its context and solution.

6

Post-Submission Strategies

Grantmaking goes slowly, but most fields of science change rapidly. The result is that thinking about and writing excellent proposals is a year-round activity. In this chapter, we first look at some useful things you can be doing while the grantmaking machinery is creaking through its paces. This will include addressing the ethics and etiquette of relationships with the granting agency while your proposal is pending.

Even for the best proposal writers, life is not an unbroken succession of triumphs. Chances are that your first efforts will be denied funding. And even when you are experienced, you will be doing well to have a funding record of more than one in two. I list ways of profiting from this fact of life and mention a few unproductive responses to avoid when you get bad news. The chapter finishes with a discussion of the ins and outs of revising and resubmitting your proposal for the next round of competition.

You've submitted your proposal well ahead of the deadline. If you submitted a printed version, you should have used a trackable delivery service, either a commercial one or the "Delivery Confirmation" option of the U.S. Postal Service, and received confirmation that it got there on time. You've taken a few days to catch up on all the things you neglected in the pre-submission flurry. Now it's time to get busy again.

The Waiting Period: Now What?

It takes at least four months, more often six, and sometimes a year or more for a proposal to work its way through all the levels of review and ranking and decision that await it. This waiting period is hardly shortened at all by electronic submissions and review software.

Parallel Submissions

Don't allow yourself to wait up to a year to hear the results of one proposal before submitting the next. Your first item of business after finishing your first proposal is to start working on the next, for either a related project (modified by your increasing sophistication in this research area and for the priorities and emphases of a different agency) or a completely different line of research.

How many of these balls you keep in the air at one time depends, certainly, on the nature of your appointment. If you are in a university or research position, you will be working on proposals nearly full time at first, advancing several lines of related research, recruiting students, and teaching a relatively light schedule so you can get your research program and lab established as quickly as possible. The opposite is true of a typical small college position where a research effort is expected as part (but only a part) of your educational responsibilities. Here, your best strategy is probably not to proliferate projects, but to keep up a steady effort to maintain grant support for one or at most two research lines. If you haven't done so as part of your hiring, you need to arrive at a clear understanding with your department chair and your academic dean as to the fraction of your time you can appropriately expect to devote to research. It is demoralizing to write a research proposal and then get raised eyebrows and push-back from the dean who is supposed to sign it for the institution.

Is there anything unethical about applying for more than one grant for closely related projects? Not at all. Your chances of being funded, even with the most excellent of proposals, are never anything like 100 percent. Forty percent or less is more like it, particularly at first, and that assumes that you have not committed any of the fatal errors discussed in Chapter Five. No one expects you to wait to be denied by one agency before you apply to another. However, there is one major ethical requirement: that you keep all agencies informed about developments in regard to the proposals at other agencies.

Primarily, this means that the second of two related proposals must list the first as "pending support" as long as the first proposal is under consideration. It is also helpful to the agency and to reviewers if in your section on current and pending proposals, you make a brief, honest statement of the degree of overlap between the two (or more) proposals. Examples might be, "Uses the same starting materials and equipment, but stresses mechanisms rather than social behaviors," or "Some overlap in the objectives of this proposal, but focuses on the characterization of surface adsorption phenomena and rates of reaction," or "Overlaps substantially with the present proposal."

Remember that two proposals on the same line of research are very likely to be sent to at least some of the same reviewers, so you can't keep it a secret in any case. It is not uncommon for those double reviewers to say things like, "This is substantially the same proposal that I reviewed for the XYZ Foundation this spring." Program officers who first learn of a parallel proposal by reading about it in a technical review will not be pleased, so you might as well at least impress people by being forthcoming.

It is not a good strategy to submit virtually identical proposals to two or more agencies. The reviewers and program officers who read them may come to view you as a "one trick pony" whose mind is fixed on a single question or kind of research. There is always some way in which two proposals in the same area can be made complementary—one stressing, for example, laboratory observation and the other a theoretical approach, or one focusing primarily on synthesis and another on reactivity.

No agency—and this is particularly true of smaller, private ones—wants to fund a project that is also likely to be funded by a major federal agency because that squanders their limited funds. One of the major tasks program officers have when managing proposals is to chase down this kind of information from PIs who did not think to include it. (PIs who do not include it give the unfortunate impression of trying to hide double funding for the same research.)

Therefore, when the first of two proposals for even somewhat overlapping research is recommended for funding, you must inform the responsible program officer of the second agency immediately. Don't be coy about this, even if it means substantially reducing your chances of funding from the second agency. You may try to make a case that later proposals are complementary to (or even independent of) the newly funded one, if that is true. But make the call. You will get credit for honesty and transparency, and there can't be too much of that in this world.

In some contexts and with some agencies, it may be possible to divide the full funding for a project between agencies. This is another good question to explore with the program officers if you are offered partial funding by the first agency to reach a decision.

Conducting Research Without Grant Money

You can't spend every waking hour teaching and writing proposals. For one thing, you may have a personal life or at least aspire to one. Besides, you actually have to do some research, if only to generate those golden preliminary results that make proposals so much more likely to be funded. Can't do that without a grant, you say? Not so!

If your job has any expectation that you will involve students in research, there will be some kind of in-house support for it:

- Released time, an agreement through which part of your regular paycheck becomes, in effect, a PI stipend for you to develop your research

- Deans' or department chairs' discretionary funds

- Compensating (usually undergraduate) research students with academic credit instead of money

- Small gifts of cash or equipment from industries, alumni groups, the Rotarians, and others

- Your own unpaid, late-night-and-weekend sweat equity

None of these is preferable to having your stuff, your students, and yourself nicely supported by external grants. But it is a way of putting data in your notebook, raising student and colleague awareness of your research, and getting articles and poster presentations into the record that will strengthen future grant proposals. Also, it will advance your thinking about the research area that you wrote up in all those pending proposals. Remember that your first efforts are less than fifty-fifty bets to get funding. When they aren't and you rewrite and resubmit, it is a very good idea to show that your thinking and the state of the project have not stood still while the months rolled by.

Keeping Up to Date

The same is true about the state of the field itself. Others are out there researching and publishing away. The work you did to get a good grasp of the existing state and likely future developments in your field will lose currency. As you finish one proposal and begin the next, all that you did to break the intellectual apron strings of your graduate or postdoctoral mentor has to be done over again. It is a continuing process.

Keep going to meetings—at your own expense if you've spent your faculty travel allowance for the year—and talking to people. Stay alive to what's going on. Read the lists of active grants in your field that are posted on the Web sites of all federal and most private agencies. The NSF site (http://www.fastlane.nsf.gov/servlet/A6RecentWeeks), for example, is updated weekly with new grants in all fields. As you do this, you are filing the information according to which of your proposals (a pending one, a rewrite, or a totally new one) it affects. This allows you to start the next version, or the next new idea, with some momentum.

This is also the time to buff up the research you did in identifying the appropriate funding agencies for your work. Possibly in your urgency to get your first proposal written and submitted, you took a short-cut for this part of the job. You knew perfectly well you were going to apply for an NSF CAREER grant, for example, and that's where your heart was. Now, you have a little more time to browse down through the NASA Research Announcements (http://research.hq.nasa.gov/allhqsearch.cfm) to see if that proposal you sent to NSF for new epitaxial superconducting coatings might also be of interest to NASA's Materials Science for Advanced Space Propulsion program (http://research.hq.nasa.gov/code_u/nra/current/NRA-01-OBPR-08/AppendixG.html). Or maybe the proposal on visual perception you sent to the National Eye Institute of NIH would find a welcome at NSF's Division of Behavioral and Cognitive Science (http://www.nsf.gov/sbe/bcs/).

In short, just about all of the preparatory work you did getting ready to write and submit your first proposal is something that needs to be done again. This is part of life as a scholar, not to mention as a writer of excellent research proposals.

Other Post-Submission Contacts with the Agency

Aside from updating information on parallel proposals that were pending at the time you wrote your current proposal, there is very little reason for you to contact the agency while your proposal is under consideration, and many reasons not to.

Do not, for example, call the program officer while your proposal is pending just to see how things are moving along, whether it has been sent out for review yet, if reviews have been returned on it and how many, and how they look. All of this information is of great interest to you, of course, but your questions about it are a pain in the neck to the program officer. Even when your tenure or reappointment review is looming and you'd love to be able to tell your department chair that your big NSF proposal is attracting "Excellent" reviews, no program officer will give you that kind of information and will only be made uncomfortable by being asked to. In fact, if a program officer does say anything positive about your proposal's chances before the review is complete and a funding recommendation is made, you know that you are talking to a rookie whose opinion is not worth passing on. Be patient.

This also goes for hot and breaking experimental results from your lab, "in review" articles accepted for publication, and other information that was not firm at the time you submitted your proposal but now has now developed in your favor. This is all good information for the next proposals, but with regard to anything now pending, the proposal you submitted is the one that is on the desks of reviewers, and the program officer will almost certainly not bother to inform them or the panel of new developments in your life. The sole exception is when parallel proposals to other agencies that you listed as "pending" are either recommended for funding or denied. That *is* useful information to your program officer, and you must be prompt about supplying it.

The Learning-from-Adversity Blues:
What to Do When Your Proposal Is Denied

The period of waiting to hear about your proposal will eventually end. And, alas, it will often end in disappointment. When an agency's funding rate is a pretty good 30 percent, then, of course, 70 percent of proposals in each round are not funded. Never mind. Batting .300 is good enough for the Major Leagues, and you're going to do better than that. So let's talk about making the best of it when you don't.

Blowing Off Steam

Nobody expects you not to be upset when your proposal is denied funding. You've put a lot of effort into writing it. At some point during the writing, if only when you go back a week later and reread it, you will have convinced yourself that it's not just an "Excellent" proposal, it's an "Outstanding" one. You've produced a piece of science that is convincingly described, cleverly attacked, crying out to be done, and bound to be funded. This self-validation is the cousin (or child) of the "wholeheartedness" I talked about toward the end of Chapter Five.

So when you are denied funding for this great proposal, you can't help getting irked. In the interest of mental health, you should blow off that steam. *Just don't do it to the program officer.* All program officers dread what could be called "the call-in week"—the period when PIs are called (or told to call) to learn their proposal's fate after a meeting of the panel or advisory board. And they dread it for two reasons.

First, it is not just politeness when agencies "regret to inform" you of the bad news. Program officers really do regret that there is not enough money to fund all the good science that they see. Almost without exception, program officers of large funding agencies were once on your side of the desk, battling for research money just as you are. They know both the elation of hearing good news and the frustration of hearing, "Your proposal was not recommended for funding." One of the things they look forward to least is that call from a PI who isn't prepared to hear bad news and responds to it with words that would not have been chosen by a cooler head.

Thus, my best advice is this: When you telephone an agency to get the result of a funding decision or when, as also happens, the program officer calls you with the news, keep that first conversation short and polite—something like, "Thank you for considering my proposal and letting me know. When may I call to discuss my reviews?" is just fine. Emotional remarks will not change the outcome and only take up your time and the program officer's.

The first telephone call is not the best time to go into detail about why your proposal was not recommended for funding. For one thing, at many agencies, the telephone lines are already burdened with other investigators waiting to get their bad or good news. But primarily you will not have all of the information you need until you have studied your technical reviews and, for agencies that supply this, the summary evaluation of the panel or program officer. I'll discuss later in this chapter how to proceed when you do have this information.

Why Writing a Denied Proposal Is Not a Waste of Time (Another Pep Talk)

Occasionally one hears grumbling that submitting a proposal that is denied funding was a waste of time. This is sometimes said, in fact, by people *before* they write a proposal, as a reason for not starting. Nonsense.

For one thing, your ideas about science will have been read and responded to by leading people in your field. In most cases, your reviews will have many valuable suggestions for improving your proposal and your thinking. (I must point out that sometimes reviewers, even distinguished ones, do not tackle the reviewing job with their full attention. And they may have misunderstood the point of the proposal you wrote. I discuss what to do about this situation later in the chapter.)

Don't worry that these peers will take advantage of their anonymity to trash you, your institution, and your ideas. I have read something like ten thousand proposal reviews in the course of my job as a program officer, and the outright hatchet jobs number fewer than ten—less than a tenth of a percent of all the reviews I looked at. Some were perfunctory things that mostly revealed that the reviewer had not carefully read the proposal. Those reviews are uniformly discounted anyhow. Most of the rest were thoughtful, constructive, short essays on the ideas advanced, respectful of the PI's thought, and containing valuable suggestions and refinements. The very best were mini-articles, pages long, with illustrative figures and bibliographies.

But even if all that were not so, even if all you ever got back from the agency was a terse "Tough luck," you would still be better off for having written the proposal. It is only through the substantial effort of all that preparatory work and through the mental discipline of organizing your thoughts—mapping the nifty colors and fascinating behaviors and great outcrops onto a set of testable hypotheses—that you advance and deepen your command of your field. It is arguable that the United States has become the scientific leader of the world as much through the institution of competitive, peer-reviewed funding as through any other single factor. Writing proposals would be good policy even if everyone had all the money they needed—or if no one did. When you have written a well-crafted and competitive proposal, you have created something important and grown in sophistication because of it. That is never a waste of time.

The Four R's of Responding to a Denial (and Only Two of Them Are Good)

Even when your proposal is denied, the worst thing you can do is to meekly (or huffily) discard it. There are constructive, and less constructive, ways to respond to denial of funding.

Reading Your Reviews

At some point, usually after you already know that your proposal was denied funding, you will be sent copies of your reviews. At some agencies, these are "sanitized" by the program officer to ensure the anonymity of the reviewer, but substantive comments about the proposal itself are not changed. Also, for most agencies, there will be a summary evaluation from the panel or the program officer. You will need to read both the reviews and the summary at least twice, probably three times.

The first time you read your reviews, it will be difficult to be objective about what they say. People are not robots, and no one can read the reviews of a denied proposal without having the words colored by the relatively fresh news that these are the reason you didn't get the grant you assuredly deserved. People see and hear what they are looking for, and tend to suppress what they are not.

Thus, we are likely to seize on positive and supportive statements with the undertone of, "Look at that! They said the proposal was good! Didn't the panel see what they said?" We tend to skip over, minimize, or mentally refute negative or critical comments, thinking, "Oh, how silly! Didn't they read the proposal?" And you may get mad all over again, this time because your proposal went unfunded because of a diabolical stew of unfair negative reviews and unfairly ignored positive ones. Again, resist the temptation to complain about this to the program officer.

Before making any calls, let yourself and the reviews cool off. The longer you are in this game, the faster that will happen. Then read them again. This time, read them as if they were talking about someone else's proposal. Objectively consider the points they raise. Cultivate the mind-set of an unbiased observer, considering the reviews as the outcome of an encounter between a text (your proposal) and a human being (the reviewer).

If the review is accompanied by a score (poor, fair, good, very good, excellent), see if the score is consistent with the content of the review. But don't be fooled by high scores; remember from Exhibit 1.1 that these scores are biased toward the high end. Reviewers, for whatever reason, are more willing to write out a negative or unenthusiastic statement about a proposal than they are to encapsulate that statement in a low score. Even so, they are likely to err on the side of politeness if they can. A review may say, "This is a good, well-written, well-thought-out proposal, but . . ." It is the *buts* that kill you.

Dealing with faint praise or no praise at all. If your proposal is denied funding, it may be because reviewers, panelists, and program officers found serious flaws in it that produced strong and specific negative criticisms. More often, and far more frustrating and difficult to deal with, your proposal may end up in the "Good" or "Very Good" pool, with a good deal of praise, not all of it ringing, and very few negative remarks. The former case is easier to deal with, so let's look at it first.

Negative criticism. Outright negative criticism can arise from at least three sources:

- There is a serious flaw in your thinking that is embodied in the proposal.

- There's nothing wrong with your thinking, but your proposal was written in such a way as to allow that impression.

- The reviewer is just flat wrong, either on scientific grounds or because he or she missed or ignored what you consider to be a clear statement in your proposal.

In the first two cases, it will take some courage on your part to admit that the criticism is valid or, at least in the second case, understandable. Make every effort to believe that before you move on to the third possibility: you are faced with either a wrong-headed review or one that is invalid. Both of these are rare, but they do happen. And if the same, or substantially the same, criticism appears in more than one of your reviews, you can pretty well bet that the problem is not with the reviewers. Either your proposal contains a serious scientific flaw, or it is written so as to give that illusion.

But while you are being objective and courageous, don't slip into discouragement. It is easy at the beginning of your career to conclude from one or two denied proposals that you just aren't cut out for a life in research. That is nonsense. Would you bet on the basis of one or two funded proposals that you are destined for a Nobel? Everyone from Einstein to Galileo to Socrates got bad reviews. The only people who never did were those who never wrote proposals.

Now read the reviews again with the proposal in front of you, and mark the sections or sentences that seem to have been the sources of the negative points in the reviews. If a review raises an issue that your proposal did not address, mark the place where (assuming you agree) you should have included it. And you will know that you are being fair to your reviews, and profiting from them, when they lead you to criticisms of your own that the reviewer didn't mention. (This is another kind of whole-heartedness.)

Faint praise. If you have a lot of "Good" reviews, and few if any "Excellents," you have a more difficult problem before you. As we have seen, there are almost never a few simple flaws, the repair of which would transform a "Good" denied proposal into a funded one. Fairly often, the reviews of middle-of-the-group proposals tend to be mildly laudatory, but with relatively unspecific and nontechnical complaints: for example, "Not sufficiently innovative," or the ubiquitous "Lacks focus." A proposal that seems to have ranked somewhere around the middle of the cohort, with mild and diffuse reviews, may well be in need of a major rewriting. I'll address this situation in the "Magic Bullet" section below.

Summary statements. Finally, if you have a summary from the panel or from the program officer, read it with the same kind of attention and respect that you are now giving to your peer reviews. Almost always it contains invaluable guidance for your resubmission in the form of statements like, "Close to funding cut-off, but needs more focus" or "Ranked relatively low; needs substantial rethinking and rewriting."

Also, because it is the program officer's special responsibility to make sure that all grants advance the agency's mission, this is the place where you will learn whether your proposal does a good job of addressing agency and program priorities. Typical statements might be, "Does not go far enough in describing role of undergraduates," or, "Good job on science, not clear enough on how this applies to leukemia [or loss of diversity, or space propulsion, or urban sociology]." Because of the program officer's role as the custodian of agency priorities, you must respect these remarks. If they are unclear, you should speak directly with the program officer about them.

Getting another perspective can be very helpful. Use the experience of colleagues who have been in this game longer than you have. More experienced colleagues, and even newcomers, in your own or closely related departments or at nearby institutions, will ordinarily be very willing to read your proposal and its reviews and help you understand how the proposal can improve by responding to the reviews. If they agree with a negative comment, you can enter into a dialogue that will help you understand the problem better than the cold and one-way communication of a written review. If they don't agree, they may be able to point out how your proposal didn't guard against a wrong idea on the reviewer's part.

Colleagues may also have personal insights into the proposal writing or reviewing process that will help you put what looks like a dismissive or hostile review into a less negative context: "Oh, that's not so bad; I bet that's so-and-so, or one of his students; people from that group always say that." Or, "Sure, but they also said you should contact somebody in Dr. Wellknown's group for advice on that technique. She's an old friend. Would you like me to call her for you?" And even, "Okay, the reviewer said your protocol lacks a control group. Where did you describe your controls?"

Refuting Bad Reviews

Reviews are written by human beings. Therefore, some, a small percentage, will be flawed, wrong-headed, or even just plain unfair. Often, this is obvious enough on the face of it that the review is discounted by the panel or program officer. This is particularly true of very short, very negative reviews. So when you call the program officer to discuss your proposal (after the cool-down and after you feel that you fully understand the criticisms and the summary advice of the program officer), you are certainly justified in asking, in a polite way, whether a particular review—the one you are sure is biased or careless or wrong—figured strongly in the panel's decision not to fund your proposal. If it did, you can explore with the program officer ways in which you can make sure that any resubmission does not go to that same reviewer. (Do not try to learn the identity of the reviewer. Occasionally, a reviewer will contact you directly, thus voluntarily waiving anonymity. There is nothing wrong with that. But if the policy of the agency is anonymous review, the program officer will adhere to it.)

Whatever answer you get, do not try refuting the review with the program officer. Remember that the program officer, while a scientist more or less in your field, is also necessarily a generalist and may not be in a position to judge the merits of detailed technical arguments. And in any case, what would you accomplish by winning the argument? You still didn't get the grant.

Reconsideration: Don't Even Ask

This is probably the place to tell you that grant programs generally commit all the money they have available at each panel meeting. There are no contingency funds for denied investigators, no matter how abominably they claim to have been treated. Thus, an agency could reconsider a negative decision on your proposal only at the cost of defunding an already funded proposal and flying in the face of the whole review process. Even if this were possible, the agency would never do it. Would you want to compete in a program in which grants might be taken away if someone else whines loudly enough? So don't even ask.

Resubmission

Resubmission is not the same as reconsideration. You are always free to resubmit the same or an improved proposal for the next round of competition. Any program officer will be just as happy to discuss this with you as he or she was to discuss your original application. The principle is the same: conscientious program officers want to be sure that every proposal submitted has a realistic chance of being funded, whether that proposal is a first submission or a resubmission.

The question before you is whether to remodel and update the structure you have already created with your first submission or to bulldoze the whole thing and rethink (first) and rewrite it (only then) from scratch. The answer will depend partly on how close you came to the funding cutoff on the first try. If you were really close, the only serious changes you need to consider are those suggested by your reviews (and by the program officer, who may help you mend any misfits with agency priorities). You will also want to take into account your own maturing grasp of your field and the fact that six months to a year have passed since you submitted the first version.

The last points are crucial. Resubmission of the same proposal without updating the literature or what has gone on in your laboratory since you sent it to the agency almost never results in funding, no matter how close you came the last time. A proposal, unlike a Stradivarius violin, is a wasting asset. Life goes on; the field advances. If you are an alert and thoughtful scientist, your thinking matures and advances over the course of a few months to a year. And so does that of your competitors. Your next proposal must be better than the last one, just to score as well as it did the first time around. Fans of Lewis Carroll will recognize a famous dictum of the Red Queen's in this: you will have to run as fast as you can just to stay in one place.

Your next submission should also show that you have read and respected the reviews of the previous one. The new proposal may not be sent to the same reviewers (agencies differ in their policies with respect to this), but it will probably go to the same program officer and many of the same panelists. The exception is when you are sure that a single criticism of a single reviewer honestly arises from a quirk or pet peeve of that one reviewer. If that's so, the panel will also recognize that and may well applaud you for not yielding to it.

Some agencies require that resubmissions of previously denied proposals explicitly address the criticisms of previous reviewers in a separate section (usually early) in the narrative. Because an example is worth a thousand words, a proposal containing responses to earlier reviewer criticisms and that was able to come out successfully on the funded side of the equation is included in the companion Web site for this book (www.josseybass.com/go/sciencegrants; see the NIH/NIMH RO1 proposal, "Neurotransmitter Function in C. *elegans*").

In cases where such sections are not required, be careful how you address previous criticisms. Statements like, "A reviewer of my previous proposal had reservations about my worker safety practices, so all students will wear respirators," invite closer scrutiny of that and related issues in the new proposal. Unless specified otherwise, just write your resubmission in such a way that student safety issues, or anything like them, could never arise.

The same goes for your response to technical criticisms. If a reviewer questioned whether step 1 of your synthetic scheme can possibly work, don't (unless you must) mention the previous review; just provide yields and structural evidence in the rewrite. That is what they were looking for in the first place. Remember my prior advice not to let questions lodge and fester unanswered in the minds of your reviewers.

During conversations with program officers about resubmission, applicants sometimes speak as if there were a magic bullet: "Just tell me what was wrong with the first version," they say, "and I'll fix it." The trouble with this approach is that there are almost never one or two minor, fixable things wrong with an otherwise fundable proposal. In fact, when that is the case, the panel will often go ahead and recommend funding anyway, with the proviso that these small, finite problems be remedied before the project begins.

Chapter Five reviewed a collection of typical flaws (lack of focus, inflated language, obsolescence, poorly defined hypotheses, lack of innovativeness, and so forth) that downgrade proposals. Notice that these and many others are general, systemic problems that apply to the entire proposal, not just to some part that can be pulled out, buffed up, and plugged back into where it was.

Furthermore, even funded proposals (as you will see in the funded examples gathered on the companion Web site for this book) show some of these flaws to some degree. Rare indeed is the proposal that is shining and free of all defects of logic, organization, and language. It is only when flaws become strong enough to tip the scale that they are cited as reasons to deny funding.

The magic bullet question is looking for an answer on the order of, "Well, most of the panel believed your language was too impenetrable, particularly in the Abstract, Impact, and Prior Work sections, while a substantial minority believed that the premises of Tasks 2 and 4 were a bit shaky. One thoughtful panelist recommends that you survey the work of the Berkeley group, particularly during the years 1999 through 2002, with an eye to how severely your work is anticipated and partially scooped by theirs. About half the panel thought you needed better instrumentation, but the other half thought that was no problem."

Program officers would need phenomenal vision, time, and memory for detail to say that kind of thing about each of hundreds of denied proposals. More likely, you are going to get a summary that hints at some or all of those factors and a more-or-less precise impression of where your proposal ranked, plus a recommendation that you rethink, address reviewer criticisms (if you're lucky, they'll tell you which ones), and resubmit.

In short, the summary statement can be diffuse, perhaps maddeningly vague, because it wasn't any one flaw that sank your proposal. It was the cumulative weight of many that pulled it below others that were more cleanly formulated and more convincingly presented. What you need to think of is not so much a magic bullet as a fire hose.

Perseverance Pays

I will close this chapter with an encouraging statistic. Agencies that keep track of the results for PIs who persevere in revising, improving, and resubmitting proposals find that their eventual funding rate is quite high, significantly higher than the overall funding rate for all applicants.

At my former agency, a review of all proposals for early-career starter grants found that over 60 percent of applicants who resubmitted were eventually funded, even though the funding rate in each competition was only 35 percent. The difference reflects the 0 percent funding rate for those who took a single denial as final and never resubmitted. This is an important message: persistence pays off.

When proposal writers were asked to provide the excerpts from funded proposals that you can read on the companion Web site, one of them said, "Well, you ought to know that a previous version of that proposal was denied. Are you sure you still want it?" Of course! That's one of the messages of this book. Take adversity in stride, respond constructively to it, and you may be providing models of funded proposals for the next edition.

Summary

Life is too short to wait six months to a year to hear from the first agency you try before you move on to the next. Use the waiting time productively by preparing updated proposals to be submitted to alternative funders, by doing "unfunded" research using resources close at hand (including your own nights and weekends), and by keeping up to date the mastery of your field you acquired while writing that first proposal.

However much you would like to lift the lid and peek inside the agency while it is churning your proposal through its review works, don't call and ask for progress reports and updates on its status. Do keep the agency informed about later submissions on the same or reasonably related topics, particularly if one of these is recommended for funding before your first agency's panel meets.

When, as must happen to everyone in the course of their career, your proposal is not funded, keep your relationship with the program officer polite and unemotional. Read your reviews and any summary statements carefully and objectively.

Don't be discouraged by initial failure. Use it as a necessary step toward writing a more competitive proposal next time, helped by the advice of your peer reviews. When you are sure you understand your reviews and summary statement, call the program officer to discuss whether minor revision and updating is justified, or whether the whole project needs to be rethought and rewritten.

7

Managing Your Grants

Some people get grants more consistently than others. Not everyone manages them in such a way, however, as to keep new grants coming in the future. Agencies don't send their money out into the world, never expecting to hear from it again. Every agency will require periodic reporting from you about what you're doing and the money you're spending to do it. This chapter reviews agency reporting requirements and ways to make them easier to meet.

A priority for every granting agency is that the research that it supports is brought to a conclusion and reported to the world. You probably want that too, but some special problems can arise.

Life is full of surprises, for good or ill. I'll describe strategies for coping with unexpected events that affect the progress of your research and how to proceed when there is a compelling reason to change the scope or direction of your research.

I'll conclude with a look at how you can use your first grants to bring in new and bigger ones, and how to keep your relationship with your first funder, and all those that come after, happy and productive.

Keeping Track of Money, Time, and People

You wouldn't think it would be a big problem to keep track of the time and money you spend on a research project or the students who are being supported by it. But depending on the clarity of your institution's monthly accounting records and given the chaos that can affect the lives of academics who are busy doing everything at once, keeping track can be a challenge. This is especially true as you start working with more than one grant at a time. The problem of keeping records of your spending and your time and percentage effort, and that of the students you are supporting, becomes important when you remember that different agencies—and even different grants from the same agency—have different priorities and different purposes. It doesn't do, and it is in fact illegal, to charge to Agency A expenses that have nothing to do with the purpose of Grant A, because they were accrued in pursuit of a different project. Yet it happens.

At most institutions, of course, you will have ample administrative help in keeping your spending straight. But you also have to watch over these things yourself. You can't count entirely on administrative personnel, who may not have a very clear idea, if any at all, of the difference between the purposes of one grant and another. Telling someone, "Oh, charge that to my grant," or even, "Charge that to the NSF grant" doesn't work. If there is the least ambiguity about which grant and which budget line you mean, there will always be more ambiguity in someone else's mind than there is in yours.

Even if your research group or your department has someone whose job is to keep track of spending under various grants, be sure that you sit down at least monthly with this person and review charges and expenditures so that you, in particular, understand where all the money has gone. You will almost always find that there are discrepancies because of your part-time bookkeeping, because you don't really understand how charges are posted to your grants, or because of a lack of clarity in communication with your administrator. It is important not to let these problems accumulate, or you may find yourself supporting your research out of your own pocket.

Because you are the one who signs the budget agreement with the agency, it will be your ultimate responsibility to be sure that you spend the money the way you said you would. Deviations from the list of approved expenditures without prior approval can cause real headaches for you, your department, and potentially your entire institution. Prior approval must come from the program officer or agency contracting officer handling your grant. Depending on your institution, you may also need prior approval from the grants and contracts office. Read your award documents to find out how to request a change.

When awards are given out, agencies make clear how often (generally annually) and in what form they require you to send them reports. These will most likely include financial reports and may also include separate reports of the students and other personnel supported by the grant. In technical progress reports, you compare what you said you would do during a given year with what you actually did, then justify (or boast about) the difference.

Financial and Personnel Reporting

This is usually done by your institution's financial office staff, who gather all the charges you made against your grant onto a clear spreadsheet or other formal report, sign it, and send it in. If you have kept in good contact with your financial office, your opinion and theirs as to what has been spent in what categories will coincide. However, check the copy of the financial report that you get to be sure that you agree with each line. This is another case of something being someone else's job but your responsibility.

If you're unfamiliar with your institution's accounting staff and their ways of doing this work, talk to them until you are confident that you and they agree about how your grant will be managed. The granting agency will review your periodic financial statements carefully, because it is responsible to its board of directors, or other authorities, for ensuring that every dollar of every grant is spent in pursuit of the agency's goals.

When the agency finds a problem, such as excessive charges in a budgeted category or spending for items not in the budget you submitted with your proposal, your program officer will call and ask you about them in a friendly but businesslike way. The budget you originally submitted is a signed promise that you will spend the grant in a certain way. In fact, the congruity between your budget and agency priorities is something that panels look at in ranking proposals for funding in the first place.

Program officers take seriously any substantial deviation between your reported spending and your approved budget. When I was an undergraduate working under a summer research grant, the agency sharply questioned our group's acquisition of a window air-conditioner for the lab. It was scientifically necessary—we needed close temperature control, and our water bath thermostats had only heaters, no chillers—but it wasn't in the original budget. It wasn't until I was a program officer, decades later, that I understood the heat on both sides of this issue.

Usually, agencies will be sympathetic and as flexible as possible in allowing deviations when you contact them in advance, but less so when they find out about them after the fact. This is one area in which it is usually easier to get permission than forgiveness.

Of course, it often happens that unforeseen developments lead you to reconsider your budget details. You may want to transfer money from Travel or a salary line (preferably your own, not your students') to Supplies and Expenses when a vital piece of equipment requires expensive repair. Usually, your program officer will be cooperative about allowing these internal transfers when you request them in advance. Don't let them show up unexplained on your financial report.

Indeed, most agencies overlook truly minor deviations. But there is nothing lamer than responding to one of those friendly calls with, "Oh, I never worry about that kind of thing; it's the department administrator's responsibility." Actually, no. It is your responsibility. Not knowing how your grant money is being spent is playing with fire.

The same is true when you subcontract to a co-PI or when you pay a stipend to a student or postdoc. Make sure that these people stay accountable for their time and effort, and produce periodic reports to you of their activities, spending, and results. Also, make sure that this accounting schedule is clear from the beginning to all parties. It can be mutually embarrassing to ask for a report and find that your co-PI is not ready and able to provide it. Besides, you'll get better mileage out of your grant dollars when all parties know that accountability is being taken seriously. And you can in turn assure the agency that you have been careful with all of its money, not just the part that went through your hands.

Technical Reports

Technical reports are entirely your job. You should be collecting reports and poster and paper abstracts from students and postdocs throughout the year. These will make your job easier. Technical reports for the granting agency are usually relatively informal (and they are not sent out for peer review), so you should be able to produce them easily. Most agencies accept copies or reprints of published articles arising from the research in at least partial fulfillment of the reporting requirement.

You should report failures, equipment breakdowns, and student disappearances as well as triumphs. In fact, if you did suffer a significant setback in your ability to do the research on schedule, it is better to say so in your annual report, along with what you are doing to overcome it, than to let the problem fester and have to report a serious shortfall later.

Exhibit 7.1 contains excerpts from a technical report. The job of this report is to assure the agency that work on the project is progressing, even if no articles have yet appeared in the literature. At this level of rigor, it is appropriate to report fairly crude quantitative results like the one-significant-figure activation energy shown in the exhibit and to refer back to previous reports. The last paragraph describes an improvisational solution to a problem that was not foreseen at the time the proposal was written. This is worth reporting because it demonstrates that the PI's thinking on this project is continuing to develop.

Prompt Publishing: The Key to New Grants

The agency's goals in making a grant are not fully met until the results of the research are shared with the whole research or educational community by means of papers presented at meetings and, more important, publication of peer-reviewed articles. Remember that publication should be as prompt as is consistent with responsible scholarship.

Standards as to what constitutes an appropriate volume and rate of publication vary greatly from one scientific area to another. When arduous fieldwork is involved, as in some areas of geology, some years may pass between the initiation of a significant piece of research and its publication. When the work is done in a laboratory that is already up and running (for example, work done by a postdoctoral fellow in an established research group), the gestation period will be much shorter. Also, guidelines with regard to inclusion of coauthors vary across disciplines. In physics or oceanography, it sometimes appears that everyone the senior author meets in the elevator or on the ship gets coauthorship. In geology, smaller author lists appear to be the norm. I mention this because these disciplinary norms will be reflected in the rate of publication that can be attributed to any one author or grant.

Because the reward structure in academic life, particularly in research universities, pulls strongly in the direction of more publications, I don't need to urge readers in that setting to publish promptly. But resist the temptation to break your publications into LPUs (least publishable units). I honestly doubt that personal

REPORT ON ACTIVITIES ASSISTED BY GRANT . . . #   20564-B2

Page   1   of   2   pages

Prepared by   Thomas R. Blackburn

Weathering studies on single crystals of Madawaska, Ontario, albite (a high-sodium plagioclase feldspar) were continued, focusing on the rate of etch pit formation and growth on the (010) face of freshly cleaved samples mounted on glass rods and immersed in HCl solutions of pH 1.5 to 3.0. . . .

Consonant with results reported in 1989, pits in the (010) face grow rapidly in the [100] and [001] directions, but slowly in the [010] direction. . . .

The rate of linear and areal growth of these . . . was measured by comparing the dimensions of SEM photographs as a function of time. For example, at pH 2.5 and 40°C, the growth rate of several pits averaged 1.4×10^{-20} mol sec^{-1}, and remeasurement of the rate for the same sample at 60° yielded an activation energy of ca. 1×10^5 J. The latter value is in rough agreement with that (119 kJ) reported for the rate of appearance of albite constituents in solution during the chemical attack of polycrystalline samples. (1) . . .

Several stereo pair photographs were acquired, and a simple, approximate, but accurate algorithm was developed for calculating the relief of features normal to the picture plane when the distance of the feature from the tilt axis is unknown. Calculations applied to a few large (ten micron) etch pits confirmed the visual impression that etch pits in the (010) plane are shallow and broad; thus that weathering rates are highly anisotropic, favoring the [001] and [100] directions over [010].

(1) Knauss, K. G., and T. J. Wolery, Geochim. Cosmochim. Acta 1986, 2481–2497 (1986).

Exhibit 7.1. Excerpts from a Technical Report

bibliographies of forty to sixty articles a year can possibly represent a creativity rate of one novel, well-formulated scientific idea per week.

Those in primarily teaching settings, however, should remember that research is an integral part of teaching and that continued research support through grants depends on a reasonably steady production of peer-reviewed research articles with undergraduate coauthors. Try to make coauthorship more than honorary for your undergraduate collaborators by first involving them in the daily planning and strategizing of the research and then assigning, critiquing, and coaching them on their part of the eventual publication. Undergraduate research participation has been the foundation of many distinguished careers in science. And it is truly impressive to program officers and potential reviewers to encounter these students at scientific meetings and find them knowledgeable and eager to share their expertise and experiences.

In any case, remember to acknowledge the support of your research by the granting agency. Many agencies have preferred wording for these acknowledgments, which will be stated in the documents you received when the grant was awarded. If you can't remember where you filed it, you can probably find the recommended wording on the agency's Web site.

What About Proprietary Information?

In Chapter One, I discussed the problem of proprietary information when someone else holds it instead of you. Now you may be on the other side of that line, particularly if your research is even partly supported by corporate money. Examples might include materials science research for which pure samples of materials that are difficult to synthesize are provided by a private research lab or geophysical research that uses preexisting exploratory data from an oil company. Research in genomics or proteomics that is partly supported by (and this includes using facilities supported by) a biotechnology company may easily lead to proprietary data.

When this applies to your research, the corporate sponsor may be able to sequester and prevent publication of your research results for a certain period of time or, in some cases, forever. Generally, this runs counter to most granting agencies' interests, not to mention your own. If you are not sure of the proprietary status of your research, you must check with your institutional administration before you submit research for publication. Usually, the sponsored programs office or a technology transfer administrator can offer guidance on the question of proprietary information. (In fact, this is something that you should be clear on when you submit the proposal; see the "Talking to Agencies" section in Chapter Two.)

Dealing with Unforeseen Circumstances

When an agency makes a grant, it expects that the PI will work diligently on the research at a rate appropriate to the institutional setting. But this would be a dull world if everything actually happened on the kind of schedule that you sign your name to when you submit the research proposal. You may get recruited tomorrow to a plush appointment at a different institution. Health or other personal issues may prevent you from carrying out your research at the pace you anticipated and promised in your proposal. Your students or postdocs may leave your group for greener and more productive pastures. For better or for worse, we cannot always count on following the paths we thought we saw stretching before us.

In particular, four questions can arise and very occasionally blow up into full-scale conflict and hard feelings if you do not take leadership in defusing them: (1) how to proceed when you leave your institution during the term of the grant or when you or your laboratory become unable, for any reason, to carry out the research proposed; (2) questions about ownership and transferability of capital equipment bought with your funds from your grant; (3) changes in the direction or emphasis of the research; or (4) inability to work on the research for a period of time.

Transferring Your Grant to a New Institution

If you change institutions during the course of your grant, you can sometimes arrange to have the grant transferred to your new institution. This is, however, by no means a sure thing or your unquestioned right. Remember that your institutional setting was part of what you presented in your proposal as evidence of your ability to carry out the work. Also, remember that the grant was made to your institution, not to you personally. If the grant in question is a programmatic one, for example, to develop a curricular innovation, it will almost certainly stay with the institution rather than follow you. If it is a research grant, it is tied more closely to you personally, and there may well be a way to move it with you.

In the latter case, if you are moving to a research-friendly position at a comparable institution, both your old institution and the granting agency should recognize that you, rather than someone else, have a unique capacity to do the research. One model that works well is for your old institution to make sure that all of your bills and encumbrances under the grant have been paid and accounted for. It then submits a final financial statement to the granting agency and refunds any unspent grant money. The agency then enters into a new grant agreement with your new institution, passes along the money returned by the old institution, and schedules any future payments under the grant to the new institution. You can see that this has to be done carefully to be sure that the old institution does not get stuck with unpaid bills and that the new institution gets all the grant money that is left.

Other agencies have their own ways of doing it or reasons for not doing it. Sometimes the grant is left at the first institution, and the PI continues under a parallel adjunct appointment in order to complete the project and graduate the students supported there.

As another complicating factor, the two institutions may well have different negotiated indirect cost rates with the granting agency. If the new institution has a higher rate, the agency will be unlikely to increase the grant or agree to the diversion of budgeted direct research costs into institutional overhead. Thus, you may need to alert both institutions to the situation and persuade your new institution to accept the lower rate.

The guiding principle in all issues is that both institutions want to retain a clean legal relationship with the granting agency. Begin the process by conferring with your department chair or dean and your Sponsored Programs Office; then call the granting agency and your new institution, and be prepared to facilitate whatever mutually agreeable arrangement may be needed to get the grant safely transferred. This can easily take several months to accomplish, so if you are going to be making charges against your grant in the meantime, be sure you have the consent of all parties involved.

There will be circumstances in which you can't expect the transfer of a research grant to be automatic. For one thing, if you are simply leaving your old institution without moving to a new academic position, you can't just take the money with you and work out of your garage or, for most agencies, an industrial lab. The same will be true if your old institution meets eligibility requirements for a grant program that your new institution does not. For example, the grant program may be restricted to nondoctoral departments, whereas your new setting has a doctoral program. In this case, it is sometimes possible, but never automatic, to work out an arrangement to continue as an adjunct appointee at your old institution for the remaining term of the grant. Any such arrangement requires the goodwill and cooperation of both institutions and the granting agency, so explore the possibility as far in advance as possible and with a can-do attitude.

If you are moving from a research-friendly position to a very different one—for example, in a secondary school or a full-time administrative position—you should probably plan to turn over the research to someone else at your old institution or simply terminate the grant and return the unspent and unencumbered money. It is virtually impossible to be a full-time dean and maintain a significant research effort. If you are transferring from an academic position to a research position at a nonprofit institution, your ability to transfer the grant will depend on the eligibility policies of the agency. Read the rules, and call the program officer.

Generally, agencies are reluctant to terminate grants. They are in the business of supporting research, not of reclaiming money. If you find yourself in any of the situations identified or a similar one, make sure that you initiate a conference with your dean or department chair on what you plan to do, followed by one with the granting agency, as soon as you are sure that you will not be continuing the research at your present institution. Exhibit 7.2 illustrates some of the kinds of decisions that agencies make in this area.

Ownership and Transferability of Capital Equipment

When PIs change institutions, it is not uncommon for minor items of hardware to be swept up in the flurry of packing and transferred with their user to the new setting. Certainly, no one worries about paper clips. Even small hand tools have an irritating and regrettable tendency to move about with their users as if they were paper clips. But when capital equipment is involved, scrupulous care is essential.

The place to begin here is to recognize that the grant that bought the equipment was made to the institution, not to you, so the equipment that you purchased belongs to the institution. As long as you are the institution's employee, this generally causes no problems. Most institutions will recognize your claim to access and

Following are three situations in which the PI requests a transfer of a grant to a new institution. Let's say that you are the program officer who receives these requests. What will your decision be? For reference, I have included the Web address of the agency in all three cases, so you can look up the regulations.

• *A PI requests transfer of an NSF CAREER grant from the University of Seattle to the University of British Columbia* (http://www.nsf.gov/pubs/2002/nsf02111/nsf02111.htm#ELIG).

This violates the eligibility requirement that holders of CAREER grants must "be employed in a tenure-track position (or tenure-track-equivalent position) as an assistant professor (or equivalent title), at an institution in the U.S., its territories, or possessions, or the Commonwealth of Puerto Rico, that awards degrees in a field supported by NSF . . . or that is a non-profit, non-degree-granting organization such as a museum, observatory, or research lab."

Does it help that the holder was eligible at the time of the award? No, since "the CAREER award must be relinquished if the principal investigator: no longer holds a CAREER-eligible appointment at a CAREER-eligible organization; transfers at any time prior to or during the duration of that award to a position that is not tenured, tenure-track, or equivalent; or transfers to an organization that is not CAREER-eligible." Because the University of British Columbia is outside the United States, it is not "CAREER-eligible."

• *The PI requests transfer of an NIH RO1 research grant from the University of Denver to the University of Paris* (http://grants1.nih.gov/grants/policy/nihgps_2001/part_iib_7.htm#_Toc504812186).

Foreign institutions are eligible to receive NIH research project grants (RO1 grants). The transfer of an initially U.S.-based grant to a foreign institution is thus possible but requires prior NIH approval.

• *A PI is moving from a small state college to MIT and requests transfer of an ACS PRF Type B grant* (http://www.chemistry.org/portal/Chemistry?PID=acsdisplay.html&DOC=prf<\\>prfgrant.html#typeb).

Type B grants are limited to faculty members of departments that do not award the doctoral degree. This transfer would be possible only if the department to which the PI is transferring does not have a doctoral degree program, an unlikely circumstance for any discipline at MIT that is likely to be funded by ACS PRF in the first place.

Exhibit 7.2. Transferring a Grant When the Principal Investigator Changes Institutions

scheduling of the equipment if you are the one who got it by writing a proposal. But what if you move to another institution? Can you take "your" mass spectrometer with you? The transmission electron microscope? The Land Rovers? That answer depends somewhat on the nature of the equipment, its cost, its age, its general usability outside the context of your research, and the institution's policies. Exhibit 7.3 poses two situations concerning capital equipment and provides guidance on how they should be handled.

Nonprofit colleges and universities may be reluctant to donate anything outright to other people, but they do want to cooperate in sharing equipment, and have no interest in hanging onto hardware that will never be used. The key, again, is to recognize up front that it is your institution that holds the grant and all the goods bought with it, not you. Most institutions that you part with on good terms will be receptive to a persuasively written proposal for a permanent loan arrangement.

Following are two contrasting situations in which a principal investigator wishes to transfer capital equipment from a former to a new institution. What would you recommend?

• *The capital equipment is a fully equipped animal colony, complete with prep room, cages, and sterilizing equipment.* This is a multipurpose facility that required a substantial investment in remodeling for its installation. Many other members of the faculty, not to mention the person who will be hired to replace you, could use it. It almost certainly stays when you go.

• *The capital equipment is a custom vacuum ultraviolet spectrophotometer interfaced to a dry box optimized for spectroscopic study of late transition metal fluorides.* This most likely will be of little use to the institution or the person who will replace you. If it stays, it may end up in the dusty basement of the department or in one of your institution's warehouses. In situations like this, you may be able to work out a transfer, purchase, or "permanent loan" of the equipment to your new institution.

Exhibit 7.3. When Might Capital Equipment Be Transferable?

Changes in Research Direction or Emphasis

Research that unfolds exactly as it was outlined in your proposal is pretty dull research. Your mind, your grasp of your field, and the state of things in that field are not put into a three-year freeze just because you get a three-year grant. In the best case, an unexpected finding in the early part of a project will put you onto a completely new line of thought and experimentation. Exhibit 7.4 explores some hypothetical situations of this sort.

Once again, your best move is to do your homework. Be sure to talk to the responsible administrators on your campus and then call your program officer *in advance*, and ask what the agency's policy is toward changes of direction. The answer you get will probably also depend on the particular program under which you have been funded. "Starter Grant" programs for young researchers (ACS PRF Type G, for example) will generally be more tolerant of change than mission-oriented projects such as programmatic grants (NSF-CCLI, for example) or disease research. If the change

---

Suppose you find that the novel nanotube bundles you were synthesizing to make a new kind of light-emitting diode become room-temperature superconductors on exposure to ozone. Shouldn't you drop everything and follow this up? Or in the worst case, what if you pick up a journal and find that someone has just made your line of research obsolete, either by scooping you on it or completely undermining its premises?

In either case, it makes better sense for you to follow a new line of research than to continue on the old one. Can you do that when your proposal amounted to a promise to carry out the old line?

Agencies vary in their tolerance for, or even encouragement of, this kind of midstream change of direction. In most cases, the tendency is to be conservative. If it were common practice to change the project after the grant starts, research funding could collapse into a chaos of bait-and-switch schemes with no accountability for anyone actually doing the research they proposed. Nevertheless, agencies have no interest in indenturing you for months or years of what has clearly become an unprofitable line of work.

Exhibit 7.4. Reacting Productively to Unexpected Developments

of direction is cleared with the agency and is substantial enough, it may call for a formal change in the title and content of the proposal that governs your project.

What Happens If You Can't Work on the Research for a Period of Time?

For any number of reasons ranging from car accidents to new parenthood to abandonment by your students, it may turn out that you are simply unable to put the research time into the project that you anticipated.

As soon as you are aware of the situation, read all agency literature that bears on it, including on its Web site, and then call your program officer and explain the problem. Almost always, the program officer can work something out with you to put matters on hold for a time, possibly including a continuation of the grant period (usually called a "no-cost time extension") beyond the originally scheduled end date.

Extensions are also useful ways to spend uncommitted funds on hand at the end of your original grant period without a lot of slapdash, last-minute spending. No-cost extensions are usually for one year at a time, with some reasonable limit, such as two or three years. If your inability to carry out the research extends beyond that, you should consider terminating the grant or handing it over to a substitute PI—either of these, of course, with the knowledge and assistance of the program officer.

Writing the Next Proposal

No later than a year before an active grant expires, you should be working on its successor. At some agencies, there is an expectation that much of the agency's budget will be committed to renewal grants—grants to PIs the agency has previously funded, so they can continue the work they have started. This tends to be particularly true of a number of federal agencies: the military, the Department of Energy, NASA, or special earmark programs. At other agencies, and particularly at the private nonprofit agencies with more limited budgets, there is no separate category for renewals or continuations, and every proposal competes on its own merits with all others, whether they come from new PIs or from currently funded ones.

When you are applying to agencies that are not dedicated to long-term research support, your new proposal has two jobs. Its most important readers, the technical peer reviewers and panelists, will not have seen your reports for the previous grant but will typically know that you had one. Thus, most new proposals in the same area should contain a summary report on the work you have already completed. This will provide a starting point from which you can lay out a case for the new effort. Remember that the case you present needs to be just as fresh and sparkling as the first-time proposals against which it will compete. Thus, balance accountability for your old grant with creativity in regard to the next one, without once sounding like, "Look, I did fine with your earlier money; just keep it coming."

In short, you will have all of the preparation and rhetorical (exposition, persuasion, and credentialing) work that I described in Chapters One and Three to do all over again, as well as giving a good account of your prior grant, all within the same page and word limits.

Finally, agencies, particularly small private ones, love to hear that you have used their grant as seed money by drawing on exciting preliminary results to get new and bigger grants from other agencies, or to serve as the basis of a budget match from the new agency. When you do that, you have, in effect, multiplied their grant by a substantial factor. Luckily, you are now in a good position to do that. If you can get one grant, you can get the next one. Go for it!

Summary

Not getting grants is tedious and discouraging. Getting them lets you in for a whole new set of chores that are pleasant by comparison. Be scrupulous in recordkeeping in regard to both resources (money, time, and personnel) and scientific progress. Be sure you understand the reporting requirements for each grant you hold and that you keep them straight. Even when expert administrators assist you with this chore, you must insist on seeing and understanding everything they report.

Your job is not finished when you've done the research and understand the science you set out to understand. You also have to be prompt and conscientious about publishing, making sure to include student coauthors and to acknowledge agency or corporate support in the form they prefer.

Unforeseen circumstances, such as a change of job, health issues, or student unreliability, can cause problems that will only get bigger if you do not share them immediately and frankly with your program officer and work with the agency to manage them.

Because the grant is made to your institution and not to you personally, all assets purchased with it belong to the institution. If you leave, however, and wish to transfer your grant to a new institution, you can sometimes arrange for the equipment to follow you on some type of purchase or "permanent loan" basis.

Science, like the people who do it, is a living, changing thing, full of surprises. One such surprise may make it desirable to change the direction of your research in mid-grant. Sometimes it is the best thing for the agency as well as for you, and you may be able to work out a change of direction in collaboration with your program officer.

Writing a new proposal to an agency that has already funded your research is just like writing the first one, except that you also have to weave into it a good account of how your research advanced with the help of the prior grant. If you have leveraged that first grant to get others, all the better.

Science is an adventure and one that for most of us requires funds to let us do what we would like. This book is just one of your tools for helping you succeed in your research and career. Best of luck!

Resource A:
Checklist for Scientific Proposal Writing

Thank you for reading all the way through this book. Here is the whole thing reduced to a checklist. If you can work your way through this list, taking into account all the exhortations, cautions, and opinions offered in Chapters One through Seven, you are entitled to congratulations . . . and a big grant!

- [ ] Make a list of research ideas and follow-ons.
    - [ ] Devise hypotheses and methods.
    - [ ] Write initial titles and abstracts; start narrative.
    - [ ] Draft initial budget.
- [ ] Research the field.
    - [ ] Go to meetings.
    - [ ] Read the literature.
    - [ ] Talk to colleagues.
    - [ ] Check research Web sites.
- [ ] Collect preliminary data.
- [ ] Research funding agencies.
    - [ ] Check your institution's agency lists.
    - [ ] Talk to sponsored programs office.
    - [ ] Read agency Web sites.
    - [ ] Talk to colleagues.

- ☐ Check award amounts.
- ☐ Pick agency short list.
- ☐ Download and read application forms.
- ☐ List questions for agency.
- ☐ Call program officer and discuss the following:
    - ☐ Eligibility
    - ☐ Fit with agency program
    - ☐ Fit with specific grant program
    - ☐ Funding rates
- ☐ Target one agency.
- ☐ Complete first draft of proposal.
    - ☐ Lay out research time line.
    - ☐ Complete narrative:
        - ☐ Introduction and bibliography
        - ☐ Hypotheses and methods
        - ☐ Impact and significance
- ☐ Begin revisions.
    - ☐ Revise abstract and title.
    - ☐ Revise narrative.
    - ☐ Give to colleagues for feedback.
    - ☐ Revise in view of feedback.
    - ☐ Build final budget.
    - ☐ Complete application.
    - ☐ Recommend reviewers.
    - ☐ Include curriculum vitae, institutional setting, and other supporting information.
- ☐ Submit to sponsored programs office for review and signatures.
    - ☐ Review before submitting to agency.
- ☐ Submit to agency.
- ☐ Take the lab to lunch.
- ☐ Start next proposal.

Resource B: Companion Web Site

This publication consists of two parts: the book you hold in your hand and a Web site (www.josseybass.com/go/sciencegrants) on which I have placed material that I believe will be useful to beginning proposal writers. These Web resources will be updated regularly to reflect the continuing evolution of the grants world. The Web site currently includes the following items:

Links to Grant Sources

"A Directory of Useful Information on Grants and Granting Agencies in Science"—thirty-three live links to funding agencies and grants clearing-house sites

Examples of Funded Proposals

1. ACS Petroleum Research Fund Type B: "SEM Stereogrammetric Studies of Etch Pit Growth Rates in Weathering Reactions"

2. NASA Mars Data Analysis Program: "Sedimentary Geochemistry of Martian Samples from the Pathfinder Mission"

3. NIH National Institute on Aging, RO3 Small Grants Program: "Cysteine Dioxygenase Transgenic Mouse Model"

4. NIH National Institute of Child Health and Human Development, RO1 Research Grant: "Spatial Analysis in Children with Focal Brain Injury"

5. NIH National Institute of Mental Health, RO1 Research Grant: "Neurotransmitter Function in C. *elegans*"

6. NSF Research Experiences for Undergraduates: "Chemistry as the Focus of an Interdisciplinary Summer Research Program at Wellesley College"

7. NSF Course, Curriculum, and Laboratory Improvement Program: "Laboratory Investigations Using Quantitative Microscopy"

8. NSF Research in Undergraduate Institutions: "Peptide Based Macrocycles as Chemosensors for Metal Ions"

9. The Research Corporation Cottrell College Science Awards: "Path Instabilities of Air Bubbles Rising in Fluids"

10. The Research Corporation Cottrell College Science Awards: "Size Effects of Barium Titanate Nanocrystals Investigated by Near-Field Scanning Optical Microscopy"

11. U.S. Department of Energy: "Controls on Gas Hydrate Formation and Dissociation, Gulf of Mexico: *In Situ* Field Study with Laboratory Characterizations of Exposed and Buried Gas Hydrates"

Resource C:
For Further Reading

There are dozens of books in print about grant writing. Almost all of them are about writing proposals for the support of community-based service projects, schools, churches, arts organizations, and the like. That is completely different from the rigorous, anonymous-peer-review world of scientific grants. The three books listed next are about proposals for scientific research.

Friedland, A., and Folt, C. *Writing Successful Science Proposals*. New Haven, Conn.: Yale University Press, 2000. Somewhat focused on the NSF application.

Locke, L., Spirduso, W., and Silverman, S. *Grants That Work: A Guide for Planning Dissertations and Grant Proposals*. (4th ed.) Thousand Oaks, Calif.: Sage, 2000. Primarily about dissertation proposals, but a good resource for graduate students.

Reif-Lehrer, L. *Grant Application Writer's Handbook*. Sudbury, Mass.: Jones and Bartlett, 1995. Strongly focused on the NIH application. Predates the formal adoption of the NIH review criteria described in Chapter Three.

A complete and valuable guide to proposal writing is provided by the University of Michigan at http://www.research.umich.edu/proposals/PWG/pwgcontents.html.

Of the many books that have been written about building a career in college and university teaching and research, the following classics are still worth reading:

Barzun, J. *Teacher in America*. New York: Doubleday, 1954.

Beveridge, W. *The Art of Scientific Investigation*. New York: Random House, 1957.

Highet, G. *The Art of Teaching*. New York: Random House, 1950.

References

Blackburn, T. R. "Non-Rationality in Science." Paper presented at the Forum on Physics and Society, American Physical Society Winter Meeting, Boston, Dec. 1971.

Blackburn, T. R. "Mercury in the Sediments of the Horwer Bucht (Lake Lucerne) and Tributary Streams, Switzerland." *Schweiz. Zeitschrift für Hydrologie* [Swiss Journal of Hydrology], 1973, *35*, 201–205.

Halpern, A. R. "The Rhetoric of the Grant Proposal." Council on Undergraduate Research Proposal Writing Institute, Juniata College, Huntingdon, Penn., July 2001.

Hammett, L. P. "Linear Free Energy Relationships in Rate and Equilibria Phenomena." *Transactions of the Faraday Society*, 1938, *34*, 156–165.

Index

44, 49–50; proposal review criteria, 50, 87;
study sections, 5; Web site, 5, 44
National Science Foundation (NSF), 23,
33, 44; Course, Curriculum, and Laboratory Improvement (CCLI) program, 32,
125; deadlines, 78–79; Experimental Program to Stimulate Competitive Research
(EPSCoR), 82; Major Research Instrumentation (MRI) program, 32; nontechnical
funding criteria, 58; Web site for new
grants, 44, 101
New money, 29

O

Obsolete research, 84
Office of Naval Research, 25
Overhead. *See* Indirect costs

P

Packard, David and Lucile, Foundation, 22
Panels, 4, 6–9, 36–37
Parallel proposals, 98–100
Patent policies, 28
Peer review. *See* Review
Pending proposals. *See* Proposals
Personnel, in budget, 63–64
Postdoctoral fellows, 33; salaries, 64
Poster presentations, 94
Preliminary results, 87, 93–94
Primarily undergraduate institutions (PUIs),
18, 32–34, 67, 98
Principal investigator (PI), 17; curriculum
vitae, 72–73; salary, 63
Program officer, 2, 27–29, 75, 102, 125, 126
Project summary. *See* Abstract
Property, intellectual, 12, 75
Proposals: appendices, 76; audiences, 36;
denied, 102–112; identical, 99; pending,
75–76, 98–100, 102; poorly organized,
87; ranking, 4–5; resubmission of, 29,
109–112; scoring, 6–9, 81–96; stages of
review, 2–6
Proprietary data, 11, 28, 119–120
Publication, 28, 117–119
Publication list. *See* Bibliography

R

Ranking of proposals, 4
Reagents, 67
Refuting reviewers, 108
Reporting: on research, 116–117; on spending,
113–116
Research: assistant, 88; change of direction of,
125–126; continuation, 29; delimiting,
12–14; "Edisonian," 85; ethical and safety
considerations in, 86; fundamental versus
applied, 30; obsolete, 84; out-of-date, 83;
status of, 10–11; undergraduate, 3, 16, 54,
73, 94, 98; unfunded, 94, 100
Research Corporation, 23, 33
Resubmission, 29, 109–112
Results, 48–49; negative, 53; preliminary, 87,
93–94
Review: anonymous, 4, 74, 105, 108; by colleagues, 55–57; criteria, 50, 58, 87; peer, 7,
105; technical, 2, 7, 36, 83, 104
Reviewers, 79; hostile, 75; recommending,
74–75; refuting, 108
Reviews, 103–108; incidence of unfair, 104;
negative, 106, 108; reading, 105
Revision of denied proposals, 109–112

S

Safety, 53–57, 86
Salaries, 25, 63–64
Samples, 52
Scores, 7; "Excellent," 92–96; "Fair," 84–86;
"Good," 86–91; "Poor," 81–83; "Very
Good," 92
Scoring and ranking: criteria for, 6; difference
between, 7; summary, 7
Search engines, 23–24
Seed money, 100, 127
Significance statement, 42, 48–50; in NIH
funding criteria, 50
Spending, keeping records of, 113–115
Sponsored Programs Information Network
(SPIN), 23, 24
Sponsored programs office, 22, 62, 73, 78
Starter grants, 18, 26, 125
Start-up funds, 73
State science boards, 25